PASSENGER STEAMERS ~ of the ~ BRISTOL CHANNEL

A Pictorial Record

NIGEL COOMBES

Three of the four Barry company steamers at their home port: the *Barry* with her 1907 funnel markings of red, white and black, the *Devonia* in the middle and the *Westonia* outside – both with red funnels and black top. Funnels of all three appear to be well scorched with hard running. All three ships passed into Campbell ownership for the 1912 season.

The funnel markings of the *Barry* have often been a point for speculation. George Owen notes that they were the same colours already in use on the tramp steamers owned by William Thomas Symonds (minus an "S" on the funnel) – Symonds was a principal of the Barry Railway Company.

The majesty and power of a paddle steamer at full speed is a wonderful sight. An extra dimension has been captured by the aerial photographer in this awe-inspiring portrait of the *Bristol Queen* taken in the 1956 or 1957 season. A large crowd is enjoying the sunshine on deck and the canvas dodger on the bridge has been rolled down, as was the custom in fine weather. Captain Jack George, Master of the *Bristol Queen* and Commodore of the Fleet, is the officer on the left.

The fine entry of the bows, the broadening beam, the powerful projection of the paddle-boxes and the magnificent, flecked wake of the paddles reaching out astern show off the *Bristol Queen* to perfection. Many have argued that she was the most beautiful paddle steamer ever built: this picture might help to confirm that opinion.

TWELVEHEADS PRESS

First published 1990 by Twelveheads Press,
Chy Mengleth, Twelveheads,
Truro, Cornwall TR4 8SN.
ISBN 0 906294 23 1

FOREWORD

MY FIRST INTRODUCTION to cruising on the Bristol Channel was when taken by my mother on the White Funnel steamer Ravenswood from Bristol to Illfracombe.

Not many years after, having made several journeys in the intervening period, White Funnel Steamers became my hobby and the interest remained with me until P. & A. Campbell Ltd. left the Bristol Channel.

As Nigel Coombes, in his admirably written book points out, the Bristol Channel is an exceptional place. Its attractive coastline, harbours, piers and resorts, open as they are to the Atlantic, make it a place of enormous charm for those who enjoy cruising.

Many books have been written about the subject but this one incorporates not only the history of passenger steamers in the Bristol Channel but many anecdotes about individuals who were concerned with the pleasure steamer industry.

It is a good thing that the history of the past has been so interestingly set out, not only for the benefit of those with nostalgic memories but also for younger generations and those who will study the subject in years to come. It is with great pleasure that I commend to the reader this excellent book.

Happily, the cruising tradition carries on and the *Waverley* and the *Balmoral* still ply the Channel thanks to the activities of the Paddle Steamer Preservation Society – long may they continue to do so.

S. Clifton Smith-Cox.
9 July 1990.

Uniform button of P. & A. Campbell.

THE BRISTOL CHANNEL
A VERY SPECIAL PLACE

THE BRISTOL CHANNEL has always been a fascinating place, quite unlike the river estuaries in any other part of our island.

For a start, it is far and away the widest of our British seaport approaches. Southampton's waterway is squeezed between the narrow confines of the mainland and the Isle of Wight. The estuaries of the Humber, Mersey and (dare I say it) the Thames are modest wedges of water leading, nonetheless, to great ports. In Scotland, the Clyde does, perhaps, offer a more spacious introduction, but much of the Firth is taken up with peninsulas and islands; in fact the Firth, by ice-age accident, is cramped for open water.

The Bristol Channel, by comparison, is wide and uncluttered, breathes unfettered sea air and, during periods of south westerly gales takes the full fury of Atlantic rollers to its margin, an unhindered onslaught which has shadowy connections 2,000 miles away on another Continent.

Significant excursion fleets of other regions were firmly based on rivers – the Thames coast, the Forth, and the Clyde – or on specific local areas such as the Solent and North Wales. The routes which these fleets plied generally embodied frequent pier calls and close coastal navigation: the atmosphere was distinctively local – cosmopolitan, perhaps, but still local.

That kind of cosy intimacy has never been a part of the Bristol Channel scenario. Magnificent paddlers, indeed, sailed down the River Avon but if you talked of "Avon passenger vessels" as you might talk of the Clyde steamers, a Bristolian would most probably imagine you were referring to the Pill Ferry (estimable as that fine little institution was). Such a difference in perception already begins to define the Channel as a place apart from other steamer strongholds.

A white funnel stands above the packed stages of Ilfracombe pier. Passengers embark from the lower level while those above watch with patient interest. It is a minor drama that never fails to attract an audience.
Alan Kittridge collection.

The tidal conditions, the variations in wind and weather, the exposed nature of the piers and the length of some of the crossings in the Channel are a constant source of surprise and interest to those unfamiliar with the area.

The tides, in particular, present a special problem. One only has to visit the stretch of the estuary near the Severn Bridge to experience it: at high water on spring tides you will see a turgid, brown expanse of water, static and unlovely. Then the ebb begins, slowly at first, gathering in intensity and speed, slicing through the uprights of the bridge, scouring the muddy bottom at up to eight knots (although twelve have been recorded at The Shoots), until six hours and forty feet later the evacuation ends. At its strongest, this great flux of water is an awesome sight. Only the Bay of Fundy in Newfoundland has a higher rise and fall.

The strong currents and admittedly lesser fluctuations, lower down, are a feature of the Channel as a whole and they do not make life easy for steamer operators. At Clevedon Pier, for example, passengers can amble down the pier to the landing stage and step up the gangway, straight onto the ship alongside, what could be simpler? That was at high water! A few hours after a cruise around the Holms the passengers return to Clevedon to find the steamer lost amongst a network of wooden piles and are confronted by five flights of steps to the top of the pier. One of the first tasks a new crew-member learns is that of propelling wheel-chairs courteously and efficiently aloft in an atmosphere of initial disbelief, growing trust and then amusement. For the less infirm amongst us it is simply to question the wisdom of that large fish supper we have just consumed.

These great tidal fluctuations (greatest at the period of the fortnightly spring tides) mean that, inevitably, some of the steamer destinations dry out altogether at low water – Minehead, Watchet, Porthcawl, Bristol – and others such as Clevedon and Penarth are untenable for shorter spells. Ilfracombe, alone, has the unrivalled advantage of being permanently accessible – a crucial factor in the continued success of passenger trips in the Channel, though even Ilfracombe is much dependent on sea and wind and unavailable at the bottom of the equinoctial spring tides. Swansea is inaccessible one hour each side of low water.

However, the ordered movements of nature have another trick to play and it is a timetable-compiler's nightmare. Since the tides' timings progress, regularly, by roughly one hour each day, the ship's roster must work on a fortnightly basis, or put more simply: a steamer passen-

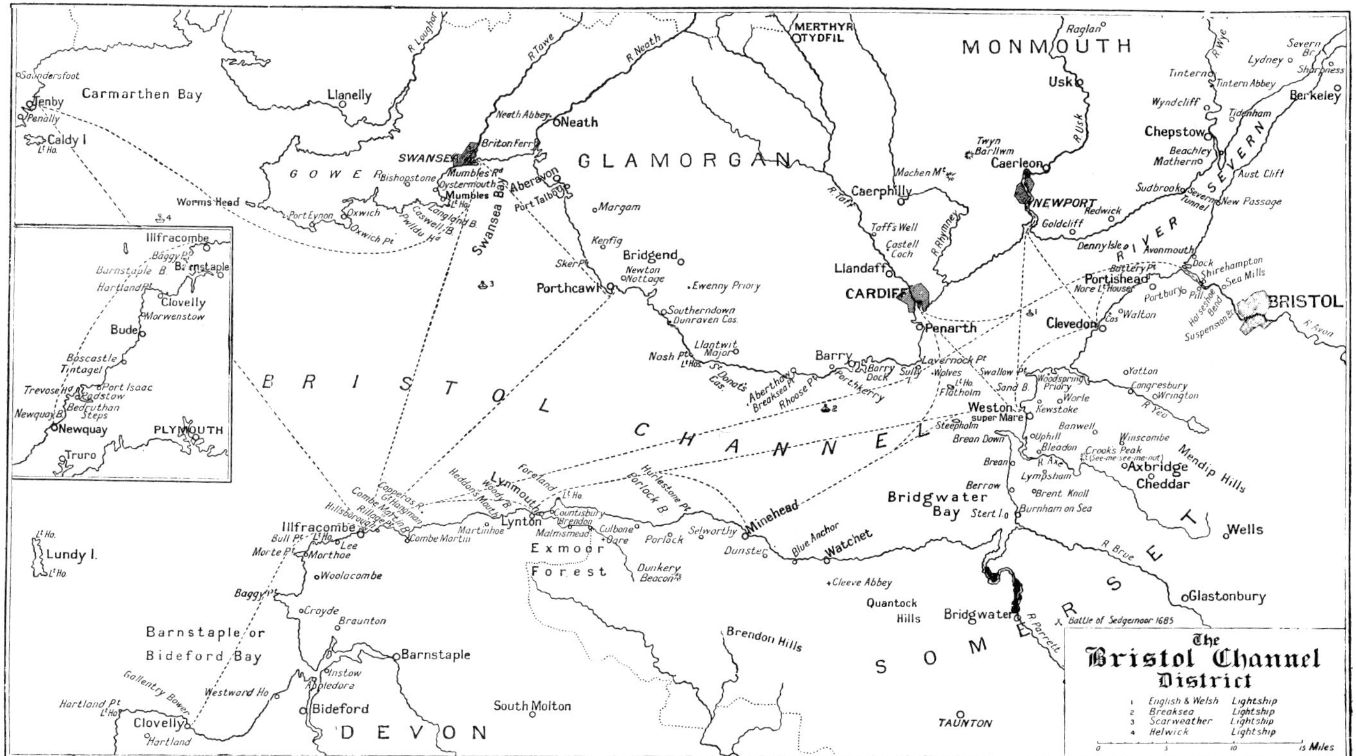

Copyright F G Warne 'Bristol Channel District Guide'

John Bartholomew & Son Ltd Edin

ger leaving Hotwells at high tide one week will find only a muddy trickle of the Avon the following week. Our steamer must ply her trade somewhere else, down the Channel. Oh, to be cruising out of Southampton with its four tides daily.

The final, truly wild card is played by the elements themselves, which call for a high standard of seamanship and ship-management skills. The deep depressions rolling in from the south-west approaches are easy to spot; a ship-master can plan for those, but the weak troughs of low pressure which could be something or nothing call for highly tuned decision-making when passengers' safety and comfort are at stake.

If an entry in the steamer's log-book reads: "Nash Point, wind S W force 6 and freshening; heavy swell", thoughts of an hour or two in Swansea recede by the minute. Sure enough, the ship's nose digs a fraction deeper into the swell as the helm goes over to port, away from the shore. She ploughs deeper still into a trough and rolls heavily beam on to wind and sea. The steamer perches and thuds, turns and lifts again with a final side-long heave and, suddenly, it's all behind her as she retraces her wake with wind and sea running astern, bidding farewell to a day's revenue. The Tannoy offers profound regret and "Time ashore at Penarth".

Who would contemplate running a steamship company on the Bristol Channel!

Happily, in the spirit of Victorian enterprise, several companies took up the challenge and forged a tradition which has spanned a hundred years – somewhat tenuously at times – to give pleasure to generations of people from both sides of the Channel and beyond.

There is a heavy swell at the entrance to Ilfracombe Harbour and despite the greyness of the day there is a handful of passengers on the promenade deck of Campbell's *P.S. Ravenswood*. Her port paddle wheel is almost completely visible but amidships a lady is not deterred from fluttering a white handkerchief to a friend on the pier. Astern, the winchman coils his rope. In this picture, dated 1909, the south-westerly surge across the harbour mouth means that the *Ravenswood* has had to turn, inside the harbour, for maximum speed to get steerage way into the elements. This is a wonderful rough weather study which required good timing from Phillipse, the Ilfracombe photographer, to gain the maximum effect of the roll.

THE EARLY YEARS

PACKETS AND TUGS

Latterday passengers travelling on the Bristol Channel can be forgiven for believing that trips from one side to the other, or down the coast, began (and even finished!) with the White Funnel Fleet. That is not surprising, given the charismatic hold that P & A Campbell's magnificent paddle steamers had – and still have – on public consciousness and affection in their unbroken association with these waters between 1888 and 1981.

The Campbells' rise and ascendancy tends to eclipse the contributions of other companies, such as the Barry Railway Co. and William H. Tucker & Co. of Cardiff. There were the red and yellow funnels of these companies respectively, at different periods in competition with Campbells before the white funnels reigned supreme.

One of the earliest paddle steamers to ply the Bristol Channel appears on this £5 note issued in 1835 by the Ilfracombe Bank. Perhaps the bank's two principals, William B. Vye and George Harris, felt that their fortunes would in future be dictated by the advent of the paddle steamer; perhaps the ship represented on the engraving was none other than the *Duke of Lancaster* which in 1822 became the first regular steamer, calling on her way from Bristol to Cork. Whatever the identity of the ship there is no mistaking the location. *Courtesy Michael Messenger.*

Among the passenger-carrying tugs was C. J. King's *Merrimac* seen here on the wall outside the locks at Hotwells. During the 1880's the *Merrimac* (96grt), built in 1883, spent part of the season on the Bristol – Chepstow run with company consorts *Sea Prince* and *Sea Queen*.

Richard Clammer collection.

Neither must we forget that steamships, propelled both by paddles and screw, had plied the Channel from the earliest days of steam machinery for some sixty years before the Campbell brothers arrived in Bristol. There is an absorbing chapter in Grahame Farr's *West Country Passenger Steamers* which outlines the first faltering attempts at providing regular services and the subsequent flourishing of the Channel packets wherever trade induced and passengers offered.

By the end of the 1820s routes had already been established from Bristol or Portishead to Ilfracombe; from Bristol to Chepstow, Newport, Ilfracombe and Tenby; and from Swansea to Bristol. By the end of the 1830s the steam packets were carrying passengers from Cardiff to Bristol or Portishead, while the Bristol steamers had extended their runs to embrace Carmarthen and Milford Haven. In the following decade Bristol was linked to Barnstaple, Bideford and Hayle, and services up-channel began from Neath, Lanelly and Port Talbot.

This is, perforce, a sketchy picture but it serves to show the variety of passenger routes in the Bristol Channel

Ilfracombe was a very different place in about 1870 when this early picture of the packet *Eclair* was taken. There is no outer pier and passengers embarked from the steps to be conveyed to the steamer by small boat as she was too big to be accommodated alongside.

The *Eclair* (210grt) built in 1865 at Port Glasgow by Kirkpatrick & McIntyre plied the Bristol Channel between 1869 and 1872 for F. J. D'Oyly & J. F. Ransom, who used her between Bristol, Bideford and Tenby on various sailings. She was originally built for the Dartmouth Steam Packet Co. Ltd. and plied from that port to the Channel Islands and St. Malo.

Alan Kittridge collection.

which were already established long before the advent of the pleasure fleets. That the early packets, in addition to their commercial role, also carried excursionists during the summer months is well evidenced.

Smaller steam vessels, sometimes owned by the packet companies, made their contribution, too, to the growing overall picture. By the 1850s, for example, Knightstone Harbour at Weston was already being used for occasional trips to Minehead, Newport and Cardiff; you could go from Watchet to Ilfracombe, and there were frequent trips to Flat Holm and Steep Holm from both sides of the Channel, largely, it is suggested, to take advantage of the welcoming tavern on each island. Bristol to Ilfracombe, however, was still a two day adventure.

Finally there were the steam tugs. These paddle vessels served a primary function of towing sailing ships, against the prevailing westerlies, right to the mouth of the Channel, clearing Lundy Island and the treacherous North Devon coast. In addition to their long-haul and docking duties, paddle tugs from all of the major ports were found to be quite suitable for summer excursion work. Their broad beam gave them more deck-room than might be imagined and from the late 1840s some of them carried passengers during the season to any suitable pier or slipway on scheduled trips.

For thirty years this was the pattern of excursion work – a fairly local and low-key arrangement, leisurely but regular, somewhat lacking in comfort and employing ships not specifically built for the job.

It merely remained for the potential delights of cruising to be exploited. There was a populous hinterland; there were, by now, unrivalled railway connections, there were suitable piers but, above all, some of the finest coastal scenery in Britain was waiting to be enjoyed and explored.

WOULD THE RAILWAYS HELP ?

Apart from a special contribution to Bristol Channel excursions by the pierhead station at Barry, the railways played surprisingly little part in the development of pleasure cruising, unlike comparable areas such as the Clyde or the Solent.

On the Clyde an increasing number of commuters found it possible – indeed desirable – to travel to the great centres of commerce and industry such as Glasgow and Paisley from their homes on islands like Bute and Cumbrae or from the Cowal shore. The railheads of Wemyss Bay, Gourock and Craigendoran entrained the steamer passengers within minutes and conveyed them efficiently to work.

Likewise in the Solent, the Isle of Wight railway network funnelled their passengers through the steamer landings at Cowes and Ryde to their mainland occupations in Southampton and Portsmouth. Such a steady, daily commuter custom on their regular services gave the steamer companies a financial cushion. This offered considerable scope for cruising more widely at other times of the day and for exploiting considerable patronage from the holiday resorts in the area. Combined rail and steamer excursions were therefore a natural and profitable option. Benefitting from their association with the railways, both the Clyde and the Solent figured prominently in the last days of regular steamer excursions.

Circumstances in the Bristol Channel area were markedly different: the capital city of Wales, Cardiff, and the chief city of the West Country, Bristol, were self-sufficient centres. Commuting from one side of the Bristol Channel to the other was not a very practical proposition anyway, the fickle tides saw to that. The Bristol Channel steamers, therefore, quickly established themselves almosy exclusively as pleasure steamers. The wonder was that the habit of pleasure cruising became so deeply ingrained that the tradition has survived right up to the present day.

If the railways were unable to fill the steamers with a host of daily commuters, would they supply huge numbers of cruise passengers?...a forlorn hope! With the exception of Barry and Portishead, at every Bristol Channel pier or port of call the railway stations were at least half a mile inland or located on the other side of town.

At least the Bristol and Portishead Pier & Railway Company chanced their luck. The company opened its line to Portishead in April 1867 and to its adjacent pier in June 1868. But it became obvious from the outset that the engineer of the project, John Francis Ranald Daniel, was far more interested in the goods traffic which would come from his Portishead Dock (opened on 24 February 1877).

In an extremely rare print from the 17 September 1870 edition of the Illustrated Midland News, a passenger train enters Portishead's first railway station (there were three and, incredibly, this building survives). Daniel operated a small number of paddle steamers from Portishead, but there was little rationale in travelling by rail from Bristol to Portishead when a steamer could be taken from Bristol itself.

THE EXCURSIONS BEGIN AND THE PHOTOGRAPHERS ARRIVE.

We are very fortunate that the arrival of the excursion ships coincided with another important phenomenon. Refinements in the art of taking pictures and the boundless diligence of the early commercial photographers handed down to us an admittedly haphazard – but utterly priceless – legacy of ship photographs, pier studies and seaside views.

PORTISHEAD PIER AND RAILWAY

This powerful study of the *Westward Ho* in her early years was taken by John York at the mouth of the Avon. The narrow white washboard below the promenade deck railings always helps to identify this steamer from her near sister, the *Cambria*.

Photo: John York; A. Kittridge collection.

In 1888, the same year in which the paddle steamer *Waverley*, Capt. Alec Campbell in command, arrived to run excursion trips in the Channel, George Eastman perfected his revolutionary, new Kodak camera. This American invention, incorporating a roll of celluloid negative, rendered obsolete the old apparatus which required glass plate negatives and ultimately gave the photographer a freedom and scope he had hitherto not enjoyed.

The classic marine photographs of the great Bristol collection of John York were originally printed from glass plates and they exist as a wonderfully comprehensive record of shipping in the River Avon. That there were so many paddle steamer plates is, indeed, a thrill for those of us who have spent happy hours in the Bristol Museum studying them.

One of York's favourite vantage points was on the Somerset side of the Avon – either on the towpath or above the railway line – exactly opposite the bottom of Bridge Valley Road. With the sun behind him (an important consideration) and split-second timing, York could capture his paddlers at speed before they slowed for the pontoon just up-river; a judiciously selected angle gave him almost the full beam of the ship. The result was a series of pictures of serene power and graceful activity amidst the rugged splendour of the Avon Gorge: a well exploited formula, perhaps, but why neglect what nature has so bountifully supplied!

The choice of pictures for a book of this kind has not been the easy task which might be imagined. The last thing I wished to present was an arbitary selection of what might be considered the best shots of ships at speed somewhere in the Channel, though there are some of this type which are so outstanding as to merit inclusion.

My aim, therefore, has been to highlight the steamers, their crews and passengers in a variety of different situations, in seascapes, at piers, in dry-dock, at pontoons and, on rare occasions (thank goodness!) even aground. Ultimately it is the day-to-day flavour of Channel life over the last hundred years which I hope to recapture in pictures which tell a story in time and place, as far as the blunting edge of time will allow. From that standpoint this is not a history book and it also explains the inclusion of some illustrations which the purist might call marginal but which, to me, spell out some of the delights of the region which passengers travelled to see.

Local photographers, like Batten and Phillipse of Ilfracombe, even met the steamers as they arrived, took pictures of groups of passengers and had postcard prints ready for sale by departure time for the visitors to take or send home.

An extremely well balanced group photograph by Batten of Ilfracombe. The passengers are aboard the *Devonia* on 5 August 1912.

Alan Kittridge collection

WHITE FUNNELS PADDLING SOUTH:
CAMPBELLS & THE COMPETITION.

IT WOULD BE quite wrong to ascribe to any particular year in the Bristol Channel or to any particular vessel the claim to be the starting point for the "Age of the Pleasure Steamers". However, the time was right for a Bristol group to form the Bristol Steam Yachting & Excursion Company to charter the Scottish paddle steamer *Bonnie Doon* in 1886. She made her first sailing on 20 May for a three hour channel trip and went to Ilfracombe and back, from Bristol, on the following day. The season was a success.

The attraction of regular excursions from Bristol during that summer, and their potential rewards, emboldened four Bristolians (George Nurse, William Galbraith, Samuel J. King and Christopher J. King – the latter two being members of the tug owning family) to buy the *Bonnie Doon* and run her in the following summer of 1887.

More importantly for the future of Bristol Channel excursions, a broadly similar group of Bristol businessmen under the title of the Bristol Channel Marine Excursion Company, chartered, that same summer, the paddle steamer *Waverley*. Under the management of her owners and with Capt. Alexander Campbell supervising, the *Waverley* ran her first trip from Bristol on 24 May and from Bristol to Ilfracombe on 25 May with civic dignitaries. From the experience gained, Capt. Alec. Campbell himself ran the *Waverley* against the *Bonnie Doon* in 1888 and the important decision was taken to set up permanent operations in the Bristol Channel in the following year.

Campbells had been established steamship owners on the Clyde for many years and Alec's brother Capt. Peter Campbell, was, meanwhile, looking after their Scottish interests from their base at Kilmun on the Holy Loch. Increasing railway competition on the Clyde left little scope for the smaller independent companies; Campbells sold their two Clyde steamers on 1 January 1889, thenceforward to concentrate on building a new future on the Bristol Channel.

The story of the *Campbells of Kilmun* by Iain Hope admirably traces the Scottish roots of the steamer family and their reasons for entering the Bristol Channel trade. Certainly the two Captains, Peter and Alec. were good seamen – Peter a qualified Chief Engineer too. They were highly professional, imaginative in their plans for the Bristol Channel and extremely determined.

They needed to be determined because they met instant and growing competition from the Cardiff excursion company of Frederick Edwards and George Robertson, who introduced their new paddle steamer, the *Lady Gwendoline*, in July 1889, as a rival to the *Waverley*. In 1890 Edwards, Robertson had also acquired the *Bonnie Doon* which ran against Campbells from August on, but the *Lady Gwendoline* must have been a bitter disapointment to the Cardiff company – high fuel consumption was the cause – and this paddler was sold at the end of the season.

There was to be no respite for the Scottish newcomers: in 1891 Edwards, Robertson introduced the handsome new paddle steamer *Lorna Doone*, on which they pinned their hopes for the Bristol service, down-channel. Campbells, however, were ready to meet the challenge with the *Ravenswood*, their first new ship built for the Bristol Channel.

The *Waverley* and the *Ravenswood* competed with the *Lorna Doone* again in the 1892 season. The *Ravenswood's* battle with the *Lorna Doone* reached a peak of notoriety the following year, when on 5 June 1893, the two ships collided while racing to be first at Weston Pier. Though there was no personal injury the *Ravenswood* was found to be at fault in the ensuing court case. In September 1893, Edwards, Robertson added an older paddle steamer, the *Scotia*, to their fleet.

Campbells were determined not to lose the initiative and had ordered a second paddle steamer, the *Westward Ho*, which began service in 1894. When the *Cambria* was delivered to Campbells in 1895 not even the brand-new *Lady Margaret* could help Edwards, Robertson. At the end of that year the Edwards, Robertson fleet (the *Bonnie Doon, Lorna Doone, Scotia* and an earlier *Lady Margaret*, which had been on the Weston run) was sold to John Gunn of Cardiff.

The new *Lady Margaret* was absorbed into the Campbell fleet in 1896, the same year in which Campbells, with a triumphant flourish, established their supremacy with the arrival of the paddle steamer *Britannia*. Of the final phase of early competition it is worth recording that the *Lorna Doone* was sold to the Southampton, Isle of Wight & South of England Royal Mail Steam Packet Company in 1898. The *Bonnie Doon* and the *Scotia* were bought by Campbells from John Gunn in March 1899.

It was an astonishing achievement. In twelve years from their first speculative arrival with the *Waverley*, Campbells had moved their operations to Bristol, registered their new company, P. & A. Campbell Ltd., Bristol, in 1893, had built four superior paddle steamers specifically for the Bristol Channel and had effectively made their own terms with the opposition.

The unique Scottish connection with the Bristol Channel. which in a different form still exists to this day with another *Waverley* and the motor ship *Balmoral*, was never cut, for Campbells always returned to Scotland for their new steamers. We must be grateful to the Campbell brothers for their vision in moving south: a hundred years of pleasure cruising has been their undying legacy to South Wales and the West Country.

The *Bonnie Doon* was the first genuine excursion steamer in the Bristol Channel arriving at Bristol for the 1886 season. She is pictured here in Cumberland Basin, Bristol. Built by T. B. Seath & Co. of Rutherglen in 1876, the *Bonnie Doon* (240grt) was an iron paddler with a single cylinder engine; she served two owners on the Clyde (with a spell at Liverpool in between) before her Bristol charter. After one season – 1889 – in Belfast Lough she returned to join Edwards, Robertson's fleet, passed to Gunn's (also of Cardiff) before joining the White Funnel Fleet at the start of the 1899 season. The years 1906-1910 saw the *Bonnie Doon* on the south coast of England but she did return to the Bristol Channel again for the next three seasons, being sold for scrap at the end of 1913.

She was a good-looking little steamer typical of the early Clyde paddlers with her long, narrow open foredeck: it is incredible to observe that despite her length of 218 feet., she was a slender wraith of a ship in the beam – a mere 19 feet. wide, excluding the paddles.

Courtesy Woodspring Museum

The *Waverley* (258grt) was built in 1885 by H. McIntyre & Company of Paisley. After the arrival of Campbells' newer ships she was put on the Cardiff to Weston service in 1894, remaining in the Bristol Channel, apart from the 1911 season when she ran on the south coast of England and also opened up operations from Hastings. During the First World War the *Waverley* served as a minesweeper from 1917 to 1919 after which she was found to be beyond repair and scrapped in 1920.

One of the disappointments of photograph collecting is to find that priceless views have been spoiled. Sometimes the ravages of time and excessive exposure to the light ruin good prints. One firm, Pincock of Bristol, suffered badly in this respect: here the fault seemed to lie in the processing of their gloss prints, which have faded badly in the course of a hundred years. It is a pleasant change to find the *Waverley*, therefore, leaving Cardiff at the turn of the century in this sharply defined Pincock print. In this case the copyist has caught the deterioration before it was too late: copying, by a good modern photographer, has saved many a view for posterity. The thrill is that so many of the photographs reproduced in this book have withstood the test of time – to say nothing of postal handling, inherent decay and the heavy-handed viewing of many generations.

George Owen collection.

An admiring crowd of onlookers has gathered at the entrance to the City Docks at Bristol for what appears to be a special occasion. The decks of the *Lorna Doone* are packed and it is a reasonable guess that this is a photograph of her maiden trip from Bristol in 1891. *The Lorna Doone* was a good-looking ship and the Campbell brothers must have viewed the new paddler with some concern.

Close to Bathurst Basin in Bristol is the curiously named Velindra public house. It commemorates one of the early Bristol packets which belonged to Pockett's of Swansea. The *Velindra* (199grt) was built in 1860 by C. J. Mare of Blackwall for the Cardiff Steam Navigation Company, running regularly on the return Cardiff to Bristol sailing. She passed to Pockett's in 1867 and for the next thirty years was sporadically to be seen "up The Cut" at Bristol.

As cruising became more popular the *Velindra* often went to Ilfraconbe, Clovelly, Lundy Island, Padstow and Tenby but her regular run was from Swansea to Ilfracombe daily. This delightful photograph (c. 1890) shows her at Clovelly. Withdrawn in 1896 the *Velindra* was broken up the following year in Clevedon Pill.

Campbells' *Westward Ho* (438grt) which was built by S. McKnight & Co., Ayr, entered the fleet in 1894, being photographed by York shortly afterwards in the River Avon. She was the first ship built for the Company with a continuous promenade deck – a sensible arrangement and of pleasing appearance, which offered increased deck space and a full length saloon below.

The *Westward Ho* had a long and successful career on the Bristol Channel, punctuated by reliefs on the South Coast in 1934 and an interesting diversion in the two preceding years, based at Plymouth and Torquay, making calls as far apart as Weymouth and Penzance.

She served as a minesweeper in both world wars but was beyond economic repair after her second spell at war. She was broken up at Newport in 1946.

A short-stay intruder in the Bristol Channel at a particularly critical time was the paddle steamer *Brodick Castle* (283grt) which appeared at Bristol in May 1894. Both Campbells and Edwards, Robertson must have been disappointed to see another vessel interrupting their attempts to gain the upper hand. The *Brodick Castle* was built in 1878 for Clyde service, went to Bournemouth in 1887 but was chartered by Bristol & Ilfraconmbe Pleasure Steamers Ltd., just for the one season. The company failed in mid-season and Edwards, Robertson took over the charter. In 1901 she started a nine-year career in the fleet of Cosens of Weymouth. Cosens found her useful, but in this picture at Bristol, dating from 1894, the *Brodick Castle* hardly looked a match for the new *Westward Ho*.

Only recently has Bristol begun to value again the variety and range of the waterfront elevations which can be so attractive in this area of the city.

A scene of shipping activity at Ilfracombe Pier in 1894, showing the motley assortment of early paddlers that were then cruising, along with the purpose-built newcomers.

At the Stonebench from the inside are the *Lorna Doone* (1881), *Westward Ho* (1894), *Brodick Castle* (1878) and the *Earl of Dunraven* (1888). The *Brighton* (1878), issuing a thin wisp of steam, and the *Ravenswood* (1891) are the paddlers at the outer berth. The inclusion of the *Brodick Castle* is interesting for she was employed in the Channel for this one season only. A former Clyde steamer (*Brodick Castle*), a passenger-carrying tug (*Earl of Dunraven*) and an ex-cross Channel packet(*Brighton*) represent a good selection of the kind of tonnage with which local owners sought to capture the market.

Campbells' *Britannia* is on the inside berth at Ilfracombe with the *Scotia* in the middle and the *Lorna Doone* on the outside as the most recent arrival. Since the *Britannia* had entered Bristol Channel service in 1896 and the *Lorna Doone* was to be sold after the 1897 season. This interesting picture of the *Scotia* and the *Lorna Doone*, under John Gunn's ownership, can be fairly accurately dated.

It must have been a perfect day, the sea flat calm at Ilfracombe, in 1895, for the photographer to have captured this unique row of seven steamers in line. From left to right are the *Ravenswood, Lorna Doone, Bonnie Doon, Cambria, Westward Ho, Earl of Dunraven* and *Alexandra* (just departing).

Campbells' ships were the *Ravenswood* (1891), *Cambria* (1895) in her first season, and *Westward Ho* (1894). The *Lorna Doone* (1891) and *Bonnie Doon* (1876) were the Edwards, Robertson representatives – they also managed the *Earl of Dunraven* (1888) for Martin and Marquand, the Cardiff tug-owners. The *Alexandra* (1854), owned by James Jones of Swansea looks, and indeed was, with her clipper bow and thin funnels, something from another age.

Built of iron as the *Aquilla* (304 grt) by Henderson & Sons, Renfrew, the *Alexandra* had run between Harwich and Antwerp, then from Weymouth (and later from Plymouth) to the Channel Islands for other owners. Her stay in the Bristol Channel was for part of two seasons before she was again sold on.

A fascinating photograph from the early competitive years, when Edwards, Robertson & Company were still battling for the Bristol trade with P. & A. Campbell, shows rival steamers at Ilfracombe.

From the inside out we have the *Velindra* (Capt. Wm. Pockett), the *Waverley* (P. & A. Campbell), the *Bonnie Doon* and the *Lady Gwendoline* (both of Edwards, Robertson & Co.).

The *Bonnie Doon* had been sold by her Bristol backers and came out in Edwards, Robertson colours (yellow funnel, red band and black top) in August 1890 to join the *Lady Gwendoline* for the tail-end of the season. The *Lady Gwendoline* (318grt) had been built by J. McArthur & Company, Paisley in 1889 but her boilers proved so unsatisfactory that she was reboilered in 1890 and then sold at the end of that season to German owners.

George Owen collection.

The lower end of the Floating Harbour at Bristol was always a hive of activity during the winter months, for it was here in this area, down Hotwell Road, at the Mardyke, by the timber yards opposite, and in Underfall Yard, that the Campbell fleet laid up, was refurbished, dry-docked and painted for the following season.

In this photograph the *Britannia* is seen during the winter of 1912-13, by the Mardyke, being fitted with new lifeboats forward of the paddle box – the davits have yet to be painted with black bases. The *Britannia* remained the flagship of the fleet from her building in 1896 until the arrival of the *Bristol Queen* in 1946.

What a wonderful sight it must have been before the 1914-18 war in this reach of Bristol Docks! White funnels everywhere, waiting in anticipation for the coming of the spring.
Photo: Stevens of Bristol; A. Kittridge collection.

An engine-room view shows the *Britannia's* shining machinery – incorporating two cylinders, therefore two cranks – all the workings open to the admiring gaze of the passengers in the alleyways on each side. Cast in top-grade steel these were the parts which never wore out.

The *Britannia* was an extremely good-looking steamer and from this angled stern view, leaving Ilfracombe with her decks filled, she was particularly impressive. The *Albion* is the other steamer, on her way in.

The *Lady Margaret* (378grt) built in 1895 by A. McMillan of Dumbarton was a good-looking paddle steamer, seen here pulling away from Ilfracombe pier. Managed by Edwards, Robertson for the one season she was bought by Campbells in March 1896 and run by them until the end of the 1902 season. She was sold to the Furness Railway Company. This picture shows her in White Funnel colours in 1896.

Courtesy Bristol Museum

The *Glen Rosa* was a pretty little steamer, but already twenty years old when Capt. Alec Campbell bought her in 1897 at the end of March. Ownerhip passed to the limited Company in 1898 when she started on the south coast of England. She returned to the Bristol Channel for six seasons before going south for a long spell between 1903 and 1912.

It seems appropriate to feature the *Glen Rosa* in a South Coast photograph taken by W. A. Pelly, Campbells' Eastbourne agent, on an extremely busy day. The *Glen Rosa* (296grt) was built in 1877 by Caird & Company, Greenock and was scrapped in 1921 being unfit for further service after the war. She was a typical Clyde steamer of her time with a long open fore deck and a bridge behind the funnel: it is a great pity that no examples of this classic Victorian genre were ever preserved. Photograph pre-1911.

The *Scotia* (260grt) was an unusual looking ship. If the forward well deck had been enclosed to gain a continuous promenade it might have given her a very pleasing profile. She was built by H. McIntyre & Company, Paisley, in 1880 for Clyde service with Capt. William Buchanan, passing to the Glasgow & South Western Railway Company in 1891 before joining Edwards, Robertson's fleet in September 1893, on the Bristol Channel.

Campbells acquired her (via John Gunn) in 1899 with other remnants of the opposition. This photograph shows her in Campbells' colours sometime before 1903 when she was bought by Italian owners.

It was a great thrill when this extremely rare photograph of the *Waverley* paddling astern from Portishead dock entrance was made available from the collection of Richard Havard.

It is of interest, too, to note that the steamer is not using the pier: moorings were easier on the stone wall outside the dock, which is the current practice with the present *Waverley* and the *Balmoral*. The picture pre-dates 1906 when the old *Formidable* (seen on the right) was towed away for scrapping.

The uncanny Scottish connection with the Bristol Channel is cemented by this modern photograph taken eighty years later of the current *Waverley* as she gently noses her way in to Portishead. The clearing mist shows the gaunt silhouettes of Royal Portbury's giant container cranes and an early morning sun breaks through the murky atmosphere to speckle the still waters of King Road. It is one of those moments of exquisite peace, broken indistinctly by the tinkling of an engine-room telegraph and the steady flap of muffled paddles, the poetry of sound and sunlight before the rigours of a busy day.

Photo: Alan Kittridge

The steamer sailing from Newport pontoon offered much to see as it passed down the River Usk: there was the Town Dock (opened 1842), almost continuous wharves and smaller shipping down the west bank culminating in the entrance to the Alexandra Dock, opened in 1875. The present lock, 1000 feet long and 100 feet wide was opened by Prince Arthur of Connaught in 1914.

Huxtable Bros. took these two pictures of the *Albion*, the first an attractive full- length study (pre-1912) and the second a longer range shot, taken after the *Albion* had passed beneath the Transporter Bridge.

Newport's Transporter Bridge is truly one of the wonders of South Wales, now shamefully neglected and disused. The latticed steel towers are 242 feet high and the central girders are 645 feet long, carrying the runners from which the platform is suspended by cables. The bridge was opened for traffic in 1906. Those who have used it will agree that it was a unique experience although the magnificent view from the central footway, reached by eleven flights of steps, was strictly for the very fit and sound of limb. That there should be a question mark over the future of this extraordinary monument to engineering skill is very sad.

The *Albion* (363grt) was acquired by Campbells in 1899. She had been launched six years earlier at the yard of J. & G. Thompson, Clydebank, for the Belfast & County Down Railway Company.

The turn of the century was a time of buoyant optimism in the Bristol Channel excursion trade. This period was also to see the sudden, even dramatic, rise in the popularity of the picture postcard. It was a social development which was to have far-reaching consequences not least as a colourful and inviting medium for advertising purposes.

A number of pleasure steamer companies took advantage of the new craze and it led to an exciting profusion of outstanding company sets. There were for example the Southampton & Isle of Wight set by Frampton, the Cosens "Flag" set, the McCorquodale set for the Glasgow & South Western Railway Company and the superb Reliable and Caledonia series in Scotland, all using advanced colour tinting techniques.

Of great interest to local collectors and historians was the "On Board..." series advertising "P. & A. Campbell's Bristol Channel Pleasure Steamers". The first set of six steamers, with very distinctive red-scrolled lettering was available by 1899/1900.

P. S. Westward Ho	On trials 1894	Clyde
P. S. Cambria	On trials 1895	Clyde
P. S. Britannia	29th June 1896	Maiden trip, arriving at Ilfracombe
P. S. Albion	1899	Arriving at Ilfracombe
P. S. Ravenswood	Late 1890s	At Penarth
P. S. Waverley	Late 1890s	At Penarth

THE GREAT DAYS
1900-1914

THE ARRIVAL OF the new century is a proper point to take stock of the Bristol Channel scene as a whole. In just a few years the excursion paddle steamers had opened up exciting new concepts in travel and trade: there was not an important resort down both the English and Welsh coastlines that did not have a regular service of some kind from those great centres of Bristol, Cardiff and Swansea.

The daily departures and evening returns gave opportunities, unrivalled even by the railways, to visit new and unusual destinations – some of them well off the beaten track – which satisfied the zest for travel. The wealthy were already embarked on their Grand Tour of Europe or summering in the popular French coastal resorts. For the less adventurous or wealthy there were comparable pleasures to discover along our own shores within easy reach and at far less cost.

"Pleasure Sailings to the Picturesque Resorts of the Bristol Channel by the Magnificent Saloon Steamers of the Bristol Channel Passenger Service Ltd.": thus ran the alluring text on the colourful cover of Edwards, Robertson & Co's weekly, pocket timetable which offered a trip (saloon return) from Bristol to Lynmouth by the *Lorna Doone* for 5 shillings (25p) in September 1891. Advertising became an increasingly important consideration.

Campbells' greatest coup in 1893 was the appearance of the first Bristol Channel District Guide, a comprehensive gazetteer of the resorts on the steamer routes and compendium of local coastal history. The updated Guide was published in numerous editions right up to the Second World War and the early editor, Arthur Levy Langfield, produced excellent work with illustrations and attractive line drawings of ships and beauty-spots to popularise the excursions.

The Guide was a superb advertising medium and the steadily increasing number of hotels and restaurants which bought space gave a clear indication of the huge benefits which the paddle steamers had brought to the flourishing holiday trade. Predictably Edwards, Robertson & Co. brought out a similar publication in 1894: the Bristol Channel Illustrated, by F. G. Warne.

Firmly established by 1900 were the summer expectations of many thousands, drawn irresistibly to the waters of the Bristol Channel from city, resort and mining community. At the piers they found comfortable, well-found steamers, on-board catering of an excellent standard and, at journey's end, a variety of pursuits to keep every family happy. And what variety, too, in the scenery! There were sandy beaches at Weston, Ilfracombe, Barry and Porthcawl; pretty Devon villages at Lynmouth and Clovelly, nestling in their private combes; the ruggedly impressive North Devon coast and the gentler bays of the Gower. They were compellingly attractive foundations on which to build an unbroken tradition. These were, truly, the great days of the paddle steamers on the Bristol Channel.

Peter and Alec Campbell might well have been content to accept the commercial success which the bounties of natural beauty and loyal custom had brought them on the Bristol Channel. Being good businessmen, however, they had seen another area which excited their interest – the south coast of England. The brothers had sent the *Britannia, Cambria* and *Westward Ho* to the south coast in 1897 for the Diamond Jubilee Review at Spithead. The *Cambria* returned to Southampton for two months later in the season and from 1900 until the war there were never fewer than two ships in South Coast waters with Brighton becoming the base for operations in 1903. The White Funnel presence there and its involvement with cross-channel excursions is most effectively covered in E. C. B. Thornton's book *South Coast Pleasure Steamers*.

The Campbell brothers privately acquired two ships from the liquidators of the Brighton, Worthing & South Coast Steam Boat Company in February 1902, bringing the *Brighton Queen* and the *Princess May* into the fold. The *Princess May* was sold in July 1902 – the briefest of stays – while the *Scotia* and the *Lady Margaret* (both ex-Edwards, Robertson paddlers) went in 1903. Then, beginning in 1905, came the epic struggle with the Barry Railway Company and their successors from whom Campbells acquired the *Devonia*, the *Westonia* and the *Barry* in December 1911. The *Westonia* was renamed *Tintern* but left for Lisbon on 2 May 1913 (having only been operated in the previous season) and the veteran *Bonnie Doon* also went in the same year.

The silencing of the opposition from rival companies had been capably executed; the weeding out of unsuitable ships made economic sense, and the introduction of three new steamers – the *Lady Ismay* (1911), the *Glen Avon* (1912) and the *Glen Usk* (1914) – placed the White Funnel Fleet at the very zenith of its popularity and influence. The diligence and flair of Peter and Alec Campbell had achieved the assembly of thirteen fine paddle steamers whose operation commanded both respect and affection: the Bristol Channel and its coastline could with all correctness be called Campbells' Kingdom. As an excursion steamship company at this stage in its history P. & A. Campbell had no rivals, anywhere.

It was an awful irony that just a few months later the sullen booming of the guns was to sound out across the whole of Europe. The animated chatter of passengers would be blotted out by the sound of exploding mines and pristine white would turn to shades of grey.

The ropemen in the bows and waist are ready to secure the *Brighton* as she makes an impressive arrival at Ilfracombe. By today's standards the turtle-back of the forecastle looks distinctly old-fashioned but from this angle, bows-on, it certainly adds to the sleekness of her profile that is lacking in other views.

The *Brighton's* dining saloon is a picture of period comfort; the upholstered settles solidly unpretentious, the buttoned backing of the continuous seating very reminiscent of the railway age, and notice the finely fretted shelving behind. The potted plants and the glazed jardiniére on the centre table are the only signs of extravagance. The atmosphere is splendidly and demonstrably Victorian. In a few moments expect a steward to stride dapperly along the carpet to collect one of the brass bells and call the first sitting for luncheon.

Some of the diners may well have been in this happy group which has just arrived aboard the *Brighton* at Ilfracombe. The fashion and varied tilt of the ladies' hats is delightful.

For three seasons Pocketts operated the *Mavis* (474grt). Originally built by J. Scott & Co., Kinghorn, in 1888 for the General Steam Navigation Company's Thames coast excursions, the *Mavis* was bought by her Bristol Channel owners in August 1910, but was withdrawn at the end of the 1913 season. Her scrapping two years later confirms that despite this quite handsome arrival at Ilfracombe (photographed by Phillipse), the *Mavis* – named in the fashion of her former owners after birds – was already past her prime mechanically. Her sisters *Halcyon, Oriole, Laverock* and *Philomel* all gave good service on the Thames.

Pockett's *Brighton* encountered brief competition on the Swansea to Ilfracombe run during the 1903-07 seasons in the very similar shape of the paddle steamer *Normandy*, owned by John Richards of Swansea.

The *Normandy* (605grt), like her competitor, had also been built by Elder & Co., on the Clyde, four years later, in 1882. The *Normandy's* original owners had also been the London, Brighton & South Coast Railway, but more immediately she had spent an unsuccessful year running between Liverpool and Douglas.

This picture, by Batten of Ilfracombe, shows her to be a better-looking ship than the *Brighton*, if only because the rake of her red and black funnels was less severe.

When the Floating Harbour at Bristol incorporated the old course of the River Avon, the New Cut was dug out to accommodate the river, which still remained tidal through Temple Meads and up to Hanham. *The Marchioness* is pictured here entering the New Cut c.1910.

There are comparatively few pictures of *The Marchioness* in this final part of her journey from Cardiff to Bathurst Wharf. The paddler's wash has reached the slipways of the Rownham Ferry which provided a convenient link from the city with Clifton Bridge railway station, on the Portishead line, and the New Inn (the white building on the left). At low water the small boats were placed abeam across the muddy trickle and passengers crossed the "pontoon" as walkers!

Photo: Harvey Barton, Bristol.

Photographs of *The Marchioness* are uncommon enough to warrant attention, particularly as seen here at her home port of Cardiff, in 1910, tied up alongside the pontoon. She was named after the Marchioness of Bute.

Richard Clammer collection.

RED FUNNEL LINE

Campbell supremacy in the Bristol Channel was seriously threatened when, in the 1905 season, the Barry Railway Company took delivery of two fine, brand-new paddle steamers, the *Gwalia* (519 grt) and the *Devonia* (520 grt). Both were built by John Brown & Co., Clydebank.

An Act of Parliament the previous year had, however, laid down stringent terms of operation: the Company was limited to running from Barry to places on the south side of the Channel between Weston and Ilfracombe inclusive all year round, and during the summer to other ports of the Channel so long as the excursions began and finished at Barry.

The two new paddlers were joined by an older steamer which was renamed the *Westonia* and put to work on the Barry to Weston ferry service.

In order to circumvent the Act, which was clearly too restrictive for financial comfort, the Company registerd the steamers under individual directors' names and adopted the management title of the Barry & Bristol Channel Steamship Company. This thin disguise was challenged in the courts with some acrimony, especially by Campbells, with the result that the Barry company was obliged in July 1907 to abide by the Act.

Before the verdict another paddle steamer, the *Barry* (497 grt), also from John Brown & Co., joined the Barry Railway Company, popularly known as the Red Funnel Line, in the same year. The company struggled on for two more seasons until in May 1910 the *Gwalia* was sold to the Furness Railway Company and re-named the *Lady Moyra*. A Cardiff syndicate, Bristol Channel Passenger Boats Ltd., took over the other three paddlers for that and the following season, but the end was close.

Campbells finally bought the *Devonia*, the *Barry* and the *Westonia* in December 1911, to hold, once more, the monopoly of the Bristol Channel excursion traffic.

The Barry & Bristol Channel Steamship Company's brand-new paddler, the *Gwalia*, her houseflag, red Prince of Wales' feathers, and name pennant fluttering at the masthead, makes a brave sight in the River Usk. Her grey hull and white upperworks show up splendidly against the Newport riverside. This colour scheme establishes the date of the picture as the *Gwalia's* maiden season of 1905, for she re-appeared the next year with black hull and red funnels.

The *Devonia* in the grey and white hull colours of her first season of 1905 enters Ilfracombe harbour. She was a very handsome ship.

The general practice was for excursion steamers to come out of Barry astern, as in this photograph of the *Gwalia*. Several pilot cutters, sails set, are at anchor in the lee of the breakwater.

The steamers, in addition to their usual services, were sometimes used on other work, as pictured in this fine Huxtable study of *P.S. Gwalia* as a troop transport. On Saturday 15 July 1905 the 4th Volunteer Battalion, South Wales Borderers embarked at Alexandra Dock, Newport, for passage to Minehead, where they attended the annual camp of the South Wales Border Volunteer Brigade.

"Such method of transport", comments Lieut. Bryn Owen, Curator of the Welch Regiment Museum at Cardiff Castle, "was quite commonplace, particularly for the 1st (Pembrokeshire) Volunteer Battalion, who embarked at Swansea during various years, bound for Minehead".

Interestingly, Liuet. Owen adds that the bush hat was widely worn by Militia and Infantry Volunteers during the period 1902-06 to allow the Army Clothing Department to clear its large stocks of bush hats left over at the end of the South African war. Some of these men had undoubtedly fought there.

Kit-bags and ammunition boxes clutter the decks and rifles have found their way on to the lifeboat cover while the army officers review the cast-off from the reserved deck. The *Gwalia's* officers, who have been exhorted by the photographer to pose at the gangway, are looking somewhat ill at ease.

There is a wet, grey sheen to the slate roofs behind the Red Funnel steamer *Gwalia*, and the passengers around the funnels look well wrapped up. The River Usk is still gently flooding towards Newport Bridge, eddying through the funnels' reflection to give a wonderful sense of movement to the static subject. The picture is dated between 1905-09.

The Barry Railway Company's new ship, the *Barry*, is seen here in 1907, her first season, arriving probably at Barry.

The *Gwalia* and the *Devonia* were the long distance paddle steamers of the Red Funnel fleet, basically on the Cardiff/Barry to Weston and Ilfracombe services. The *Barry* tended to serve Weston and Bristol from the South Wales ports, while the *Westonia* plied between Barry-Cardiff and Weston on the ferry run.

The *Westonia* (333grt) had been built in 1899 by J. Scott & Company, Kinghorn, coming to the Red Funnel steamers fourth-hand – *Tantallon Castle, Sussex Belle* and *Rhos Colwyn* were her previous names. In 1911 Campbells bought her and she sailed for the 1912 season, only, as the *Tintern* before being sold to Lisbon owners. She looked somewhat old fashioned with her square ports and open foredeck but she suited her Barry owner's needs at the time. Here she is seen leaving Weston Pier.

White table cloths and aspidistra set off this comfortable interior of the *Barry's* dining-saloon with three stewards in attendance.

The *Barry* impressively fills the West Wall of Watchet Harbour. It is her first visit, on 24 July 1907 to the West Somerset port.
Courtesy Bristol Museum.

The lay-out at Barry was very compact. Once inside the shelter of the breakwater the steamers berthed at the immediately accessible pontoon without hindrance from other commercial shipping which entered the dock to the left of the picture. There was a covered walkway, attached to the pontoon, leading to the trains for Cardiff and the Welsh valley branch lines. The whole operation was extremely convenient and bears a strong similarity to the Clyde railheads with their connecting steamers. Nowhere else on the Bristol Channel was there such a handy arrangement and it should have been a continuing success had it not been for the legal fiasco.

In this picture of 1908/9 vintage, by Heber Shervington, of Barry, the *Gwalia* is at the pontoon and the incoming steamer is the *Barry*.

This stern view could almost have been taken on the same day as the previous photograph. The *Gwalia's* decks are packed and there is an air of expectancy on board. The stern repeater, which gives the instructions to the ropeman, is on stand-by, while the ropeman looks anxiously towards the Captain on the wing of the bridge. The *Barry* is the outside steamer.

The Barry Railway Company's steamship operation had ceased by 1910. The packed decks of the *Devonia* on 14 July 1910 should have pleased her new owners, the Bristol Channel Passenger Boats Ltd., who had taken her over the previous month. However, it was to the White Funnel Fleet that the *Devonia* passed in 1911 – as graceful a ship as they ever acquired. The Bristol Channel's loss between 1923 and 1932 was the south coast of England's gain, since that is where the *Devonia* spent much of her time after the war.

Courtesy Bristol Museum.

A second set of four postcards was published in 1902, though the negatives themselves often originated as the steamers entered service.

There were also reprints, in black script, of the *Albion* and the *Britannia* cards from the first set, published 1899/1900 – the latter retitled "On Board P. & A. Campbell's Saloon Steamer, *Britannia*".

So few of these surviving postcards have been postally used that it would seem that passengers tended to hoard them as mementos of a special day out.

P. S. Westward Ho (unnamed)	Early 1900s	In the River Avon
P. S. Glen Rosa	1899	Arriving Ilfracombe
P. S. Ravenswood	1899	Leaving Ilfracombe
P. S. Brighton Queen	June 1902	Clyde, trials

Few paddle steamers had a shorter career on the Bristol Channel than the little *Princess May*. The Campbell brothers bought her from the liquidators of the Brighton, Worthing & South Coast Steam Boat Company at the same time as the *Brighton Queen* at the end of February 1902. She was repainted in White Funnel colours, spent a few weeks in the Bristol Channel in 1902, was despatched to the Spithead Review at the end of June and was sold without returning to Bristol.

Built on the Clyde in 1893 by Barclay. Curle & Company the *Princess May* (260grt) already looked old fashioned beyond her years, pictured here in Cumberland Basin – a gawky mallard rather than a white swan!

Grahame Farr collection

The *Bonnie Doon* was acquired by Campbells in March 1899 from their erstwhile competitor John Gunn of Cardiff. It was, perhaps, fitting that she should start her White Funnel career on the Bristol Channel if only as a token of her historic links in opening up the Bristol excursion trade. For five seasons (1906-1910) she spent part of her time on the south coast of England, where W. A. Pelly captures her steaming away from Eastbourne pier. The lady in the first skiff, holding her hat, and the gentleman in the second, waving a handkerchief are in for an uncomfortable few minutes very shortly! Like the *Glen Rosa*, the *Bonnie Doon* exhibits the same classic Victorian charm and poise.

Not only the residents of the Nautical School, Portishead, but also the Ashford family, of Nore Farm, looked to the bottom of their respective properties with disbelief on the foggy morning of 1 April 1907. There, on the foreshore, fast aground was the Campbell paddle steamer, *Albion*, on an unscheduled visit.

The *Albion* had left Newport in a thick fog (apparently grounding momentarily as she left the River Usk), rounded the English and Welsh Grounds, made for the Somerset shore and then lost her way as she made for Portishead and Bristol. Mercifully Capt. James Webber grounded on just about the only stretch of flat shingle in the vicinity. If it had been any further downstream the ridged outcrops of rock would have seriously damaged the *Albion's* bottom; fifty yards upstream the *Albion* would have piled into Black Nore Point.

There are several photographs of the stranding and as it was Easter Monday a steady stream of sightseers came down to have a look. The Portishead cabbies, by all accounts, had a good day transporting some of the 450 passengers to the railway station for the half-hour journey to Bristol, to view the splendours of the Avon Gorge from the railway carriage along the bank rather than from the saloon in mid-stream.

The local photographer, first on the scene, was able to capture the gusts of unwanted, exhausted steam before the boiler could be banked down, which more than adequately makes up for the mist which seems to have receded.

Judging by the tell-tale seaweed line the *Albion* grounded almost exactly on high water which explains why there was little opportunity to refloat her on that tide. None the worse for her adventure she was brought off on the next tide aided, it was reported, by seven tugs as the Nautical School clock struck nine in the evening.

Heavy weather photographs of pleasure steamers at sea are rare. Sensible passengers stayed below and few had cameras anyway. Campbells had advertised a positioning run for *Cambria* on 31 August 1908 from the Bristol Channel to Brighton – out by steamer and return by rail – it was a regular feature of the timetable, though the *Cambria's* trip to act as a consort for the *Brighton Queen* at that stage of the season was a bit unusual: Mr. Owen is of the opinion that the move was to "run off" the *Lady Rowena* then running out of Brighton.

The voyage was a memorable one. The *Cambria* left Clovelly, her last port of call, in perfectly calm weather but after rounding Land's End things went from bad to worse with heavy seas off the Cornwall and South Devon coast but worse was to come.

Off Portland Bill the *Cambria* suddenly came upon a cargo of pit-props which had broken loose from a steamer which had just foundered in the mountainous seas. The Captain and the Mate studied the danger with concern. Timber baulks in the paddle boxes could cause untold damage to the wheels. Worst fears were realised, there were pit-props in the port paddle box. The port paddle box was split open in an effort to free the wheel.

The severity of the gale is all too obvious from this shot looking forward as the *Cambria* lurches down into another deep trough. Several of the saloon windows were smashed, the forward side seats were washed overboard and a doctor on board was the first passenger injured, with a broken arm.

It was later reported that four ships had foundered in the English Channel during the storm.

It was a testing voyage and the passengers must have been mightily relieved when the *Cambria* arrived at Southampton for their calm return journey by railway train. The experience persuaded the company that for safety's sake smaller round ports must take the place of windows.

These pictures are amongst the most dramatic records of rough-sea coastal cruising ever taken and compare with Richard Danielson's pictures of the *Waverley* arriving at Douglas in a storm in 1985. They demonstrate the fine seagoing qualities of the paddle steamer in rough weather and remind us that ships smaller than these crossed the Atlantic in weather that was often a great deal worse.

Richard Clammer collection, courtesy Bristol Museum.

The Ilfracombe photographer's caption reads "Presentation to Capt. P.S. Cambria". The date of 10 October 1908 is fairly late in the season and since the *Cambria* had spent the previous month on the South Coast, following her epic voyage around Land's End, this is almost certainly the happy sequel as grateful passengers present a framed testimonial to Capt. Livingstone (directly behind the kneeling steward) in this delightfully informal grouping.

The arrival of the *Ravenswood* at Penarth shows her with two funnels. The date must be pre-1909 when she was reboilered and re-engined by Barclay, Curle Ltd., after which she possessed a single funnel. Heavy silting at the entrance to Cardiff Docks requires constant dredging: the dredger in the background is just one of those features which help to locate an unnamed view.

George Owen collection.

From the mists of time a dimly remembered occurrence was the stranding of the *Ravenswood* at Lavernock Point, in 1911. Fortunately, once again, there was a camera on hand to record the event. The picture helps to illustrate the characteristic very shallow draught of paddle steamers, when measured against the height of the onlookers on the beach.

The white funnel in this picture by Garratt (of the phantom gulls) belongs to Campbells' *Glen Rosa*. Some way behind, the *Barry*, of the Red Funnel Fleet follows her down the Avon.

For two reasons this picture is interesting. Firstly, by comparison with the present day, how wide and gently shelving the banks appear! There is no doubt that the regular passage of shipping helped to scour the sides of the heavy accretions of silt which come up with every tide. Now that the traffic is less frequent the mud-banks have steepened and solidified, bound by thick tufts of deeply rooted marram grass – that curious import designed, it is said, to prevent Braunton Burrows in North Devon from blowing away. The grass has done its job well in the last half-century. It has colonised the entire Bristol Channel and turned the littoral brown into green, a quite remarkable natural phenomenon.

A second feature is the accurate dating of the picture as 1911. Just after Easter in that year the *Glen Rosa* was fitted with round portholes and a new, white saloon (visible behind the paddle-box); the yard took great pride in completing all the work by Whitsun. In September of that year the Barry company was absorbed by Campbells and the *Barry* had lost her red funnel by the following season.

The Ailsa Shipbuilding Company of Troon (note the Glasgow & South Western Railway Company's trucks in the background) held an excellent record in the building of paddle steamers and passenger ferries. Fitting out in July 1911, was Campbells' latest ship, the *Lady Ismay* (495grt).

She proved a very useful addition to Campbells' fleet and was the first of three similar pre-war paddlers built by the Ailsa Shipbuilding Co. The *Lady Ismay's* career was a short one: called up for war service in December 1914, as a minesweeper, she was mined and sunk on 21 December 1915, off the Longsand Light Vessel.

Courtesy Bristol Museum

Avonmouth had always had aspirations to being a great liner port. Regular passenger services by the Imperial Direct West India Mail Service already ran to the Caribbean, but it was considered a notable coup for the Port of Bristol when, from May 1910, Canadian Northern Steamships Ltd. decided to use Avonmouth for its route to Montreal, and to Halifax during the winter when the St. Lawrence was closed by ice. The imposing *Royal Edward* (1907; 11,117grt) and *Royal George* (1907; 11,146grt) must have presented an awe-inspiring sight to excursionists in the Channel.

Sadly these fine ships, though calling on war service in 1915, never returned. The *Royal Edward* was torpedoed by U-14 off Kandeliusa in the Aegean Sea on 13 August 1915, with the most appalling loss of 935 lives, 860 of them soldiers (though there are conflicting figures on this point). The *Royal George* survived the war and spent two years until mid-1920 on Cunard's trans-Atlantic service, before being laid up on the River Fal and scrapped two years later.

This photograph of the *Royal Edward* emphasises the fine lines of these Clyde-built Fairfield Ship Building Co. vessels with their buff funnels topped by a blue band. As C.J. King's tug, the *Bristolian*, casts off, hundreds of emigrants line the after railings of the *Royal Edward* for a final look at the Old World, while in the distance a Campbell steamer would, no doubt, poignantly echo the fond farewells from her own decks.

It was one of the less usual roles of Campbells' paddlers to ferry ashore liner passengers from the *Royal Edward* and the *Royal George*, anchored to await the tide in King Road or Walton Bay. Perhaps the trans-Atlantic mails would also be sent early on their way; at any rate, the knot of expectant port officials would be able to deal with some of the immigration formalities before the liner docked with the heavier baggage. The liner tender in this instance is the *Barry*.

Two superb studies of Campbell's *Devonia* in dry-dock at Bristol for pre-season painting and overhaul. The dry-dock, oddly shaped and a tight fit for the *Devonia*, was owned by Stotherts, who were also iron-founders and shipbuilders, later famous for their hydraulic cranes. In the stern view, Merchant's Dock – still in existence – can be seen on the right, while the bow shot looks back towards Clifton Heights. The chippers and painters are at work below the water-line of the hull. She still carries a Cardiff registration from her years as a Barry Railway steamer. Campbell's acquired the *Devonia* in December 1911. The addition of lifeboats forward suggest 1913-14 as a reasonable guess for a date.

It is perhaps surprising that cards and photographs of steamer interiors, given the comfortable ambience of the public rooms, are found so infrequently. These two views of the dining saloon and lounge of the *Devonia* are, therefore, an important record of pre-'20s style.

Eating on board was one of the pleasures of travelling in the Bristol Channel and Campbells' catering had a high reputation. A 1920 price list offered a full breakfast (with fish, ham and eggs) for 4 shillings (20p), luncheon (with soup, fish and cold meat) for 3 shillings and sixpence (17p) and full high tea (fish, cold salmon, meat, toast etc.) for 4 shillings. It was not unusual when there were evening departures, for home-going office workers to take the sea route to Portishead, Clevedon or Weston: a wonderful opportunity to indulge in the famed cold salmon buffet after a hard day. Out by train in the morning and home in style and comfort!

The *Devonia's* dining saloon with white napery, vases of flowers on the tables and a full silver service was an impressive sight. In this picture the comparative orderliness suggests that the photographer is at work before sailing-time or on an off-service day. The lounge is comfortably and solidly Edwardian, floors and woodwork beautifully polished, and divided into private alcoves. The Ionic columns are not only ornament but help to conceal the steel stanchions supporting the deck above. In both saloons, as in many ships, mirrors are an important device to give the illusion of infinite space. On a domestic note: to the very end, all of the *Devonia's* cutlery and sugar-bowls carried the insignia of the Barry Steamship Company.

The Red Funnel steamer *Westonia* was bought by Campbells in December 1911, and by combining the uptakes had one funnel removed during the winter. Although Campbells must have intended to retain her after going to all that trouble, she stayed in the fleet as the *Tintern* for the one season of 1912, during which this picture was taken just outside the locks at Bristol. Once again the company transformed rather a dowdy duckling into quite an elegant swan. A Portuguese company bought her in May 1913.
Richard Clammer collection.

Half a dozen local youngsters have been charmingly and carefully posed (bribed ?) to wait in the foreground as the *Glen Avon* paddles down the River Usk. Huxtable Bros. photographed this in about 1913. Campbells would have been well pleased with the degree of patronage from Newport that this picture evidences.

The *Glen Avon* (509grt) and the *Glen Usk* (524grt) were very similar in appearance, the former arriving in the Bristol Channel during 1912 and the latter in 1914. Both paddlers were registered in Newport and they were the last ships built for Campbells (by the Ailsa Ship Building Co., Troon) before the First World War. Berthed here at the Newport pontoon, *Glen Avon* and *Glen Usk* make an attractive pair of almost identical twins during their respective first seasons. Note, too, the professionalism of the local photographer, W. Clifton, who has spent anxious moments waiting for a tram to complete each picture. George Owen tells me that it was an extra-special day for the *Glen Avon* – her maiden trip.

With a fine surge of paddle-wash Campbell's *P.S. Britannia* pulls away from Lynmouth after disembarking her passengers into small boats for an afternoon ashore. The Bristol photographer, Garratt, was well out of his area: perhaps he had come down-channel on *Britannia*, commandeered a boat and set up his equipment, taken this evocative picture from almost under the *Britannia's* graceful counter stern and then waited with the other passengers for the steamer to return from Ilfracombe for the trip back to Bristol. Amusingly, the photographer, feeling that his study lacked life, has added half a dozen seagulls, a feature in many of his maritime photographs. This picture predates 1914.

The Campbell brothers personally acquired from the liquidators of the Brighton, Worthing & South Coast Steam Boat Co. on 26 February 1902, two very different ships: the diminutive paddler *Princess May* and the larger *Brighton Queen*, the latter being sold to the limited company in 1903. The *Brighton Queen*, built in 1897 by the Clydebank Engineering and Shipbuilding Co., Glasgow, was refitted by Campbells, principally in the enclosing of the main deck forward of the funnel. The row of portholes, set against the white livery, emphasised the graceful lines of the steamer and was copied in all ships subsequently built for the company.

The *Brighton Queen* (553grt) was employed on the Brighton station from 1902 to 1914; however, she would return to Bristol for winter overhaul. Her immaculate appearance in this Avon Gorge picture – even down to the pristine white paint on the anchor chain – suggests her departure for the South Coast at the start of the season. Like most of her consorts she became a minesweeper at the outbreak of war and, sadly, sank off the Belgian coast on 6 October 1915 after striking a mine.

The officers and crew of the *Bonnie Doon* pose in front of the bridge for the camera during the paddler's south coast season. Capt. Hector McFadyen, in the centre, sports an R.N.R. Lieutenant's two rings, the Mate at his right, Chief and Second Engineer to his left. The seamen are in front and two of the stokers at the rear – the other two, no doubt, below making steam. The harpist and fiddler entertained the passengers during the cruise, while the two ladies looked after the needs of the lady passengers. To complete this domestic scene, stewards and purser stand behind the Captain and can you spot the two ship's kittens ? This group photograph was taken between 1906 and 1910.

Regular master of the *Cambria* for the seasons 1905 to 1914 and 1920 to 1926 inclusive was Capt. Allan Livingstone, perching on the rail, with the purser, Mr. Skinner, in front of him and Mr. Fred Nunn, the Mate, leaning on the rail. Beside the Captain sits Mr. Fred. W. Birmingham, the Campbell agent at Ilfracombe.

The Campbells' Agents were a vital part of the organisation. The Birmingham Family held a record which was unequalled in the Bristol Channel. Prior to 1897 Jack Birmingham was their Ilfracombe Agent. In that year, after a spell at Weston, Jack's brother, Frederick William Birmingham took over the Ilfracombe office in the Strand until 1941. Fred's son, Frederick Parker Birmingham, joined the organisation in 1930, succeeded his father and served Campbells until they finished in 1980.

THE FIRST WORLD WAR

THE PADDLE STEAMERS of the White Funnel Fleet were quickly requisitioned by the Admiralty when war broke out. In a conversion programme between September 1914 and July 1915, all but three ships had minesweeping gear installed and served from various coastal bases; the last two were called up in May 1917. One ship, the *Barry*, started as a Bristol Channel patrol vessel but finished as the most travelled of them all.

A Campbells' fleet list at this juncture is a most helpful guide to the ships in service prior to hostilities, gives an indication of when they were called up and when they were returned, thus setting the scene for post-war operations. In numerical terms – thirteen ships – it marked the operational peak of the White Funnel Fleet.

Name	Left Bristol	Returned Bristol	War name
A. Devonia	30 Sept. 1914	2 May 1919	
A. Brighton Queen	30 Sept. 1914	Sunk 6 Oct. 1915	
A. Cambria	2 Dec. 1914	4 May 1919	Cambridge
A. Westward Ho	2 Dec. 1914	1 April 1919	Western Queen/Westhope
A. Glen Avon	2 Dec. 1914	11 March 1919	
A. Lady Ismay	2 Dec. 1914	Sunk 21 Dec. 1915	
B. Britannia	5 Feb. 1915	8 April 1919 (Glasgow)	
B. Glen Usk	5 Feb. 1919	11 April 1919 (Glasgow)	
C. Ravenswood	1 July 1915	5 March 1919	
C. Albion	1 July 1915	15 Feb. 1919	Albyn
Barry	24 July 1915	27 Sept. 1919(Troon)	Barryfield
D. Waverley	30 May 1917	17 March 1919	Way
D. Glen Rosa	30 May 1917	23 March 1919	

The first six paddlers (A) made up the Grimsby unit. In August 1915 they swept a clear passage for the bombardment of Zeebrugge, but their minesweeping operations extended variously from the mouth of the Humber to the Belgian coast: it was here, off Nieuport, that the *Brighton Queen* struck a mine and sank with the loss of seven lives. The *Lady Ismay* met the same fate off the Longsand lightvessel. The unit, less the two casualties, transferred to the Tyne in 1917 for the duration.

The *Devonia* has arrived at Ilfracombe and it is 7 July 1914, in this Batten picture. There are summer smiles and cotton frocks and two short months hence the world will never be the same.

One cannot collect pictures like this and be immune from a heightened sense of history. How many of these young men in the camera's view made it back from the holocaust to their friendly terrace at Ashton Vale or the leafy calm of Llandaff village. Fervently we hope that many did, were reunited with the girls at their side...and five years, much agony, later, paddled down to Ilfracombe, and perhaps the Devonia would take them there...perhaps.

The second unit (B), the *Britannia* and the *Glen Usk*, was based on the Clyde. The two paddlers took up minesweeping duties here and off the Northern Ireland coast, sometimes as far as Barra Head in the Outer Hebrides. On 12 August 1915 they left Troon for the Moray Firth on the east coast of Scotland and finally finished their task based on Granton. The *Glen Usk*, with a party of 1,000 Edinburgh schoolboys aboard, made a special excursion from Granton on 21 November 1918, to watch the German fleet pass up off the mouth of the Forth to surrender at Scapa Flow.

The next unit (C), the *Ravenswood* and the *Albion*, spent the whole period minesweeping from the base at Dover, but were also present with their sister paddlers at the raid on Zeebrugge. By a trick of wartime necessity the two Campbell ships found themselves serving alongside the *Lorna Doone*, their erstwhile Bristol Channel competitor.

The final pair (D), the *Waverley* and the *Glen Rosa*, stayed very much in home waters, sweeping the mouth of the Bristol Channel between Swansea and Ilfracombe from their base at the Welsh port. Later they transferred to the Thames.

The one remaining Campbells' paddle steamer, the *Barry*, had the most adventurous war of them all. She found herself involved in the Dardanelles campaign, ferrying troops to the beach-heads established at Suvla and Anzac Bay, having arrived at Mudros on 8 August 1915. It was an ill-conceived campaign, little short of a disaster, and five months later the *Barry*, again under fire, was busy evacuating the unfortunate troops. In December 1916 she helped at the Salonika landings, spending the rest of the war in the eastern Mediterranean. After such a gruelling experience, the *Barry* was the last of the White Funnel Fleet to de-requisition.

It will be seen that the Admiralty did not hand their minesweepers back immediately. Stray mines had to be cleared and minefields dispersed long after hostilities had finished.

When the steamers returned it was found that the *Albion*, the *Waverley* and the *Glen Rosa* were in such a poor state that they were sent for scrap (the *Waverley* lingering until 1921) – a sad end for three old favourites – and two fine ships, the *Brighton Queen* and the *Lady Ismay* had been sunk.

The dark clouds of war, however, had lifted at last and it was a cheering sight for the Bristol Channel when, in 1919, it was back to peacetime pleasures for the *Cambria*, the *Glen Avon* and the *Glen Usk*.

Attentively some of the officers and Ministry personnel study the minesweeping gear which has been fitted on the stern of the *Devonia*. There are also naval officers on board, the rear saloon windows have been plated in and the paddler has donned wartime grey for the four years which lie ahead.

The date is 30 September 1914. It must have been strange to paddle away into unknown waters at the time of year when the excursion fleet would, under normal circumstances,have been thinking of laying up, here at home, in Bristol, for the winter and the crews looking forward to a well earned rest.

From Cumberland Basin, into the river lock and out into the Avon it would have been stranger still for the *Devonia* to keep well out past the landing stages. No waiting passengers for the *Devonia* today, no children for Weston sands, no happy, smiling faces for the camera at Ilfracombe. Britain was at war with Germany and the *Devonia* was answering duty's call.

This is a picture tinged with sadness. Still looking immaculate in her unaccustomed grey paint, the *Brighton Queen* has entered Cumberland Basin for the last time: it is 30 September 1914 and she is one of the first two Bristol Channel steamers away. A white funnel in the far distance has yet to be converted. Wives and friends have been invited aboard, George Owen remarked, for the short trip from Campbells' yard, through the locks, to be set down at Hotwells Pontoon before departure – the *Brighton Queen's* last passengers but with a White Ensign fluttering at the stern.

When the last two Campbell paddlers went to war another small steamer was chartered for the Weston ferry service. In this rare photograph taken from Birnbeck we see the *Duchess of Devonshire* (221grt) arriving in her stop-gap role of the 1917-18 and part of the 1919 period. She was built in 1891 by R. & H. Green, Blackwall and was owned at the time by the Devon Dock, Pier & Steamship Company Ltd. to whom she subsequently returned. The *Duchess'* normal peacetime itinerary included calls at south Devon and Dorset beaches and it was Sidmouth beach which finally claimed her on 27 August 1934, where she was wrecked.

Alan Kittridge collection.

At the end of the war the Bristol Channel minesweeping flotilla was still kept busy. Here, from the outside in at Ilfracombe are the *Waverley, Glen Rosa, Harlequin, Monarchy* and *Glen Cross*. The two Campbell paddlers are recognised from their funnel cowls. The date is January 1919.

The *Waverley* returns to her home port in wartime grey; with a hint of irony her funnel could almost be a dirty white but she was not to wear her old colours again, being scrapped in 1921.

Courtesy Bristol Museum

THE STORY OF the Bristol Channel paddle steamers after the acquisition of Tucker's two Yellow Funnel ships in July 1922 became, thencefoward, the story of P. & A. Campbell Ltd. Numerically the fleet reached its post-war strength of eleven ships, including the *Glen Gower* which ran on her maiden trip on 2 June 1922, from Bristol to Ilfracombe: she was the only new ship built for the Bristol Channel excursion trade between the wars.

There were two changes of name within the fleet: the *Barry* was renamed the *Waverley* in March 1925 and the *Lady Moyra* became the *Brighton Queen* in 1933 – both new names were more suitable for their south coast of England spheres of service, where Campbells were still heavily involved.

This profitable and successful period in its history was, sadly, to see the death of the White Funnel Fleet's two founders: Capt. Alec. Campbell died at Kilmun on 10 December 1928, aged 66 and Capt. Peter Campbell died on 18 December 1938 aged 80. It was, perhaps, fortuitous that neither lived to witness the awful depredations made upon the fleet of handsome and popular paddle steamers which they had so painstakingly built up.

Captain Alec. Campbell. Co-founder of P. & A. Campbell Ltd. He lived at Ferngrove, Kilmun, a village on the Holy Loch but spent the excursion season in Bristol. *Courtesy: Bristol Museum.*

Captain Peter and Mrs. Campbell. Co-founder of P. & A. Campbell Ltd. He lived in Clifton, Bristol until 1936, retiring to Wotton-under-Edge, Gloucestershire, after the death of his wife, to live with his daughter and son-in-law, Mr. and Mrs. Gordon Fry. *Courtesy: Anne Miller.*

What variety the paddle steamers could offer the Bristol Channel resorts in those inter-war years! White Funnels starting from Bristol, Cardiff, Newport and, now, Swansea through to the end of September. There were short channel "daps" – the Bristolians' unique expression for a minor outing! – round the English and Welsh Grounds Lightvessel, or the long day trip to Ilfracombe from any of the main bases, sometimes involving changes of steamer at Weston or Cardiff. Further down channel there was a regular outreach to Clovelly, Lundy Island and the Gower Coast – a marvellous criss-crossing of white paddle steamer wakes which reached its most concentrated pattern on the Weston to Cardiff ferry where there were up to eight crossings at week-ends.

Inevitably the slump of the early 30s, which saw shipping laid up in every major port and river estuary in Britain, must have dented Campbells' passenger returns. If there had been thought of a rebuilding programme it could simply be shelved, for though the White Funnel Fleet was ageing by modern standards it had been well maintained at the company's Bristol yard. The occasional reboilering was all that was required. Some observers have suggested that a newer fleet might have helped the company subsequently but the irony is that age is no handicap to the paddle steamer: cruising custom and changing habits were usually the final arbiter.

One curious feature of this period is the noticeable decline in the photographic record of the paddlers, for no apparent reason: here there was certainly a change. Postcards of this period were no longer the rich source of archive material that they had once been. Partly, one supposes, the novelty was wearing off; more pertinently the dwindling army of those "in service" hastened the decline. With hindsight what was sorely missed was that wonderfully diverse torrent of cryptic postcard correspondence – often nautical – which kept domestic servants in touch with the world outside when their "families" went afloat. Who would have guessed that as service became unfashionable or unaffordable our steamers would be less well documented ?

Little had changed at Ilfracombe since pre-war days: the company kiosks, the store, the bandstand and even the lines of horse-drawn vehicles on the pier were comfortingly in their usual places. There is the one hint that the year is 1920 or 1921: Campbells' *Britannia* has the twin-funnelled *Lady Moyra* astern of her, in Tucker's colours. It is a tranquil scene, all the passengers ashore and a bevy of small boats pottering about on their leisurely business.

Alan Kittridge collection

If Campbells had imagined that they would have the Bristol Channel to themselves after the First World War they were wrong. The Cardiff tug-owner W. H. Tucker bought two paddle steamers from the Furness Railway Company and began excursion services in 1919. One of the steamers was the *Lady Evelyn* but, for the larger of the two, the *Lady Moyra*, it meant a return to the Bristol Channel of the White Funnel *Devonia's* equally graceful sister, the *Gwalia*, from the old Red Funnel days.

The Yellow Funnel Line, as it was known, was augmented by a chartered twin-screw steamer, the *Robina*, late in 1919, to serve Cardiff, Weston and Bristol (also, unusually, during the winter). Tuckers, however, found the Campbell opposition too tough for them. After three seasons both the *Lady Evelyn* and the *Lady Moyra* were absorbed into the White Funnel Fleet. It had been a roundabout route for the *Lady Moyra* – from Cardiff to Bristol via Fleetwood and Barrow – but at last she was back on the Bristol Channel, where she belonged, beside the *Devonia*.

Two photographs of the *Lady Moyra* from the Tucker period taken by Phillipse of Ilfracombe are of interest. At Easter in 1919 she appeared with all-buff funnels with thin black bands at the rim. Readers can decide for themselves whether they prefer the 1921 version when blue funnel tops had been adopted and where her name has been freshly painted on the bows and also two extra lifeboats have been added in front of the paddle boxes.

The *Lady Moyra* passed to Campbells in July 1922 and, by the time Vickery photographed her the following season, she looked truly magnificent. Some would say that the famous White Funnel livery has never been surpassed by any other paddle steamer operator and this picture supports that view. The light green boot-topping just above the water, the extra thin line on the hull, the outlining of the portholes and the addition of the black cowls atop the gleaming white funnels transform an attractive ship into a beautiful one.

The sisters are re-united at Clovelly during the summer of 1920 or 1921. It is an attractive setting as the camera takes a bird's-eye view of Tucker's *Lady Moyra* just inshore of Campbell's *Devonia*. Later, when the *Lady Moyra* adopted Campbell's colours, only the presence of a single ventilator cowl beside the *Devonia's* foremast could tell the sisters apart from a distance and on the promenade deck aft, the *Lady Moyra* had two cowl ventilators abreast while the *Devonia* had one circular ventilator.

The officers of the *Lady Moyra* assembled for this informally cheerful group, between cruises, at Ilfracombe, c. 1928-29.

Back row: Mr. Burnett, 3rd Engineer; Reg Tyrell, Chief Steward and son of Campbells' Catering Manager; Mr. Meredith, Wireless Operator.

Front row seated: Mr. Jimmy Martin, Mate; Captain Riddell; Johnny Black, Chief Engineer; name unknown, 2nd Engineer; Bob Vickery, Purser.

Tuckers' other paddle steamer was the *Lady Evelyn* (320grt) built in 1900 by J. Scott & Company, Kinghorn. Like the *Lady Moyra*, she too was an ex-Furness Railway ship and also passed to Campbells for the 1922 season.

The rocky carbuncle in the distance identifies the location: Denny Island is between Portishead and the Welsh shore. This picture must date from 1920-21.

Grahame Farr collection.

Campbells' only ship newly built for them in the inter-war years was the *Glen Gower* (553grt) seen here fitting out in 1922 at the Troon yard of the Ailsa Shipbuilding Company. Another piece of White Funnel history went with her, in a very literal sense, when she began sailings that season: the engines of the old *Albion* were in such good condition when the steamer was scrapped after the war that they were re-assembled in the *Glen Gower* at Troon.

She was employed in the Bristol Channel until 1934 (except for a month on south coast of England relief in 1926) but worked on the south coast between 1934 and 1939. After the second world war the *Glen Gower* put in short spells away from the Bristol Channel in 1947, 1950 and 1953; at last, after much political wrangling the No Passport trips to Boulogne began again and for three seasons 1954-1956 she was again down south.

The *Glen Gower's* last season of 1957 was at home in the Bristol Channel, where she will be happily remembered as the Swansea steamer in her early and later years. She languished at Penarth until 7 April 1960 when she was towed to Antwerp for breaking-up.

Courtesy Bristol Museum.

The *Westward Ho* has just arrived at Ilfracombe on Saturday 29 July 1922. The passengers, well muffled for this summer afternoon, have made an effort to smile for the photographer. The Purser (two rings), Alec Taylor, judges where best to line up the gangway between the uprights of Ilfracombe Pier, while the crewman awaits the order to put the gangway ashore.

It would be true to say that P. & A. Campbell had a remarkable record for safety, but thick fog caught *P. S. Cambria* unawares on 12 July 1926, when she went aground on Rillage Point, just out of Ilfracombe.

In this photograph the *Cambria's* bows are well and truly on the rocks, but happily the ship's boats and the Ilfracombe lifeboat, which is at the stern, were able to ferry the passengers ashore. Capt. Livingstone was able to bring his ship off on the next tide. One of the ship's boats has already been hoisted aboard, while the other port-side boat is waiting in the falls forward of the paddle-box.

To add an intimately North Devon touch, a local sailing ketch bound up-channel, looms out of the mist and passes serenely by.

A handsome study of the *Brighton Queen* arriving at Brighton in 1933 for the first time with her new name, changed by Campbells from *Lady Moyra*, better to suit her south coast station. Flags fluttering in a brisk breeze, the *Brighton Queen* makes a picture of purposeful elegance.

The *Britannia* (459grt) was built in 1896 by S. McKnight & Company of Ayr and the compound machinery was made by Hutson & Sons, Glasgow. It is a contemporary curiosity that a hundred years after her launch the remains of the repair slipway next to which she was built are still visible behind a brick wall adjacent to the Ayr Yacht Club. She was reboilered in 1912, 1921, 1935 (when a larger, very handsome new funnel was fitted) and 1948 (when two new funnels accompanied her change to oil fuel).

Apart from visits to Naval Reviews on the south coast of England and seasons down there between 1948 and 1950, the *Britannia* was very much a Bristol Channel paddle steamer and an extremely popular one at that. She was withdrawn at the end of 1956 and went for breaking at John Cashmore & Co., Newport. The *Britannia's* 60-year career was exceeded only by the *Ravenswood.*

In the middle of her career, this photograph (c. 1930) demonstrates the popularity of Bristol Channel cruises: the *Britannia* surges off down the Avon, her deck densely packed from stem to stern.

The *Ravenswood* (391grt) was built in 1891 by S. McKnight at Ayr and her first engine with single cylinder came from Hutson & Corbett of Glasgow. She was reboilered in 1909 when she adopted one funnel instead of two. After the Second World War her profile altered radically with the substitution of two heavy rectangular paddle boxes. The *Ravenswood* looks much more dignified with the curved fan-light boxes in this shot from 14 July 1935: she is engaged in the intricate river turn outside the locks at Hotwells, her exhaust valve snorting powerfully.

Though she served on the south coast of England from 1912 to 1914 and from 1923 to 1925 it is as a Bristol Channel steamer that the *Ravenswood* will primarily be remembered and most specifically as the Newport ship latterly until her withdrawal for scrapping in 1955.

Courtesy Bristol Museum

The *Devonia* in her White Funnel days is working up a good speed having just left the Cardiff pontoon.

Grahame Farr collection.

This study of the *Brighton Belle* shows her to be an exceptionally pretty steamer after Campbells had made alterations to her. Formerly Tucker's *Lady Evelyn*, she appeared, more suitably named, on the south coast of England between 1923 and 1935.

She is pictured off Avonmouth in 1938 by which time horizontal vents had become featured on the paddle boxes.

Which pier is the *Glen Avon* arriving at ? Clevedon seems to be a good guess. Anyhow, the date is 22 July 1938 and the *Glen Avon* (509grt) built by the Ailsa Shipbuilding Company in 1912, is seen at her best fron this angle. She spent all of her time on the Bristol Channel and was a popular ship on the shorter ferry runs. During the Second World War her anchors fouled an underwater obstruction during a gale and she sank on 3 September 1944, off the mouth of the River Seine.

Courtesy Bristol Museum.

P.S. WESTWARD HO

P.S. Glen Avon.

P.S. Glen Gower

P.S. BRITANNIA.

P.S. "DEVONIA."

P.S. RAVENSWOOD

P.S. Glen Usk.

P.S. Brighton Queen.

P.S. Cambria.

P.S. Brighton Belle

P.S. Waverley

The period between the two world wars was one of great stability in the composition of the White Funnel Fleet. Postcard collectors and paddle steamer enthusiasts alike have found it a reasonably simple matter to make up the fine composite set of P. & A. Campbell cards from this era.

Jackson's Faces of Weston were the official company photographer at the time (1936) and though the postcards lack a sense of occasion they are splendidly sharp: a thoroughly workmanlike job and ideal for what they were intended.

The pictures of the *Britannia, Cambria, Devonia* and *Waverley* were taken at Porthcawl and exhibit a finer sense of perspective than the others, which all appear to have been taken at Birnbeck, Weston. The eleven cards are an essential document of Bristol Channel steamer history.

Courtesy Lionel Vaughan and Nigel Coombes collection.

PADDLERS AT WAR, AGAIN

THE CRY OF havoc throughout Europe again, in 1939, unslipp'd the dogs of war and, again, the Bristol Channel paddle steamers were called up to serve their country. It was a grimmer conflict for the ships of the White Funnel Fleet than the first had been and five steamers were lost, three of them in the exceptional circumstances of the Dunkirk evacuation.

Initially the steamers were grouped into two minesweeping flotillas. The 7th Flotilla, based at Granton on the Forth comprised the *Brighton Queen, Britannia, Cambria, Devonia* and *Westward Ho*. The 8th Flotilla, based at North Shields on the Tyne comprised the *Brighton Belle, Glen Avon, Glen Gower, Glen Usk* and *Waverley*.

The *Ravenswood* was brought out on the Weston ferry run for the 1940 season but by July it was considered too hazardous and she was laid up until, a year later, she was sent off to Belfast as an anti-aircraft ship. She ended the war at Devonport as a coastal trooper in the area.

The *Empress Queen* was requisitioned in 1940, after completion, and following some service on the Thames as an anti-aircraft ship she was a troop carrier from November 1943 to October 1946 on the Stranraer to Larne run.

It was at Dunkirk, however, during Operation Dynamo that the Bristol Channel paddle steamers really came into their own to perform the task, in the exigencies of war, for which they were best equipped. When the call went out on 24 May 1940, the Campbell steamers of the two minesweeping flotillas (with the exception of the *Britannia* and the *Glen Usk*) answered the request to help evacuate the hard-pressed British Expeditionary Force.

The great heroism exhibited by the officers and crews of the ships which rescued an army at Dunkirk is part of the rich maritime history in which our islands abound. Dunkirk was a deliverance of epic proportions, on such a scale that it almost defies the imagination. It was the strangest paradox that the pleasure steamers should find themselves at the very centre of the grimmest embarkation of their careers and. amidst the scenes of fearful chaos and destruction, the paddle steamers of the White Funnel Fleet ferried home their most unlikely passengers.

But at what cost...The *Devonia* was run aground and broke her back, the *Brighton Queen* was sunk by air attack and the *Brighton Belle* sank after striking a wreck. By one of those unusual tricks which time can play with old ships there is still a surviving link with the dying moments of the *Brighton Belle*: the paddle steamer which went to rescue her crew, the *Medway Queen*, lies today in a creek off her home river awaiting funds to be fully restored as a living memorial to the Little Ships of Dunkirk.

Later, the *Waverley* and the *Glen Avon* were also lost: Campbells' war record was both proud and painful.

Name	War Name	Saved at Dunkirk	Fate
Brighton Belle		800	sunk 28/5/40
Brighton Queen		700	sunk 1/6/40
Britannia	Skiddaw	not present	post-war fleet
Cambria	Plinlimmon	900	scrapped
Devonia		present	beached 31/5/40
Empress Queen	Queen Eagle (1940-43)	not present	post-war fleet
Glen Avon		888	sunk 3/9/44
Glen Gower	Glenmore	1,235	post-war fleet
Glen Usk		not present	post-war fleet
Ravenswood	Ringtail (briefly)	not present	post-war fleet
Waverley	Snaefell	981	sunk 5/7/41
Westward Ho		1,686	scrapped.

The *Britannia*, bristling fore and aft with anti-aircraft guns and carrying the White Ensign as *HMS Skiddaw*, still manages to capture the elegance of her proper calling.

Grahame Farr collection

The *Cambria* was converted into an anti-aircraft vessel in 1942 as part of the Thames defences. Here she is seen on the Thames in a quieter moment. Ministry of Defence records prove that the *Cambria* did not miss the Dunkirk evacuation (despite earlier doubts) and later finished her war service as an accommodation ship at Harwich. Sadly neglected, she was broken up at Grays, Essex, in 1946 as being beyond repair. Her wartime name as J. 66 was *Plinlimmon*.

Courtesy Bristol Museum.

The Campbells' turbine *Empress Queen* was renamed *Queen Eagle* during the first part of her wartime career, reverting to her original name in 1943 as a troop carrier. With her sides plated up she gave the impression of a very much larger ship.

Grahame Farr collection.

Two paddle steamers which returned to serve the White Funnel Fleet after the war were the *Britannia* (left) and the *Glen Usk* (right). seen here at Underfall Yard on 21 October 1945, the *Britannia* (war name *Skiddaw*) had been back in Bristol since 17 May and is still in her dazzle paint. The *Glen Usk* (war number 4-377) had arrived eighteen days before the picture was taken.

Richard Clammer collection.

PADDLE STEAMERS IN DECLINE

Big ships in the Centre at Bristol are always a reminder that the city's historic debt – for her power and influence – is to the sea. The *Empress Queen*, shown to the public for the first time in June 1947, is open for visitors to make their own assessment of Campbells' newest arrival.

Grahame Farr collection.

Campbell's returned to peacetime service in 1946, initially with the *Ravenswood*, the *Britannia* and the *Glen Usk*, in the Bristol Channel. The bright hopes for the future of paddle steamer excursions are captured in this photograph taken on 4 April 1946 of the launch of the flag-bedecked *Bristol Queen* from Charles Hill's Albion Dockyard in Bristol. The *Britannia,* her decks lined with well-wishers for this happy occasion, is still being fitted out behind the moving hull of the *Bristol Queen*, which was to be ready for service by September.

After a few weeks in service the *Britannia* suffered boiler trouble and she did not resume until 1948, reboilered and, as this picture of 9 June 1948 shows, with two handsome oval funnels, smoking well in the City Docks, Bristol. Cranes, factory chimneys, masts and funnels are capped by the slender spire of St. Mary, Redcliffe beyond.

The *Britannia* then then spent that season and those of 1949 and 1950 on the south coast of England.

Richard Clammer collection.

The *Bristol Queen* makes a fine sight as she rounds the Horseshoe Bend at speed on her way down the Avon, past the white Powder House and on to the wooded side of Hung Road in the far distance. The Powder House – rather derelict, these days – is a historic building; sailing ships bound up-river were obliged to offload any explosives being carried aboard so that, in the event of a fire within the confines of the City Harbour, there would be no danger of the adjacent warehouses going up with the ship.

The *Bristol Queen*, pictured from the air in this June 1949 photograph, is just a few minutes short of Ilfracombe, journey's end.

The narrow inlet in the background is Watermouth Harbour...and thereby hangs a curious tale. Whenever steamer people gather, rumour and speculation are ever constant shipmates: it's all part of the game to predict what will or will not happen to the ships. A well founded rumour can establish the purveyor's instant credibility for the whole season while a spurious claim can always be attributed to someone else in the first instance before subsiding into hearsay of the vaguest provenance.

One year it was confidently predicted that one, or both, of the Queens would be laid up in Watermouth Harbour for the winter; the rumoured possibility has persisted ever since. What actually transpired was this.

At the time, Roland Wickenden of Townsend Ferries owned Watermouth Harbour and Caves and with an eye, no doubt, to business asked Clifton Smith-Cox, "Could we get a ship in ?" "Yes Roland, of course you can get a ship into Watermouth Harbour," the Campbells' Managing Director asserted, but adding dryly, "The problem is could you ever get it out!"

It is very sad to see this classic gem of Bristol Channel lore laid to rest after such a gallant run.

Alan Kittridge collection.

The last-minute activity at Fairfield's yard on the Clyde heralds the imminent departure of the *Cardiff Queen*. Soon the builders' work will be done and their last paddle steamer will be ready to leave for the Bristol Channel. This 1947 photograph shows it to be a moment for quiet conferences, final licks of paint and stowing the mattresses for the bunks below. A large passenger liner is fitting out in the background.

Nick James collection.

The *Cardiff Queen* (765grt) was built by the Fairfield Ship Building & Engineering Co. Ltd. on the Clyde and after her launch on 25 February 1947 was swiftly completed and ready in June for Bristol Channel service. She was the last seagoing excursion paddle steamer to be built in Britain and, apart from the 1952-53 seasons on the south coast of England, she spent all of her time in home waters. The *Cardiff Queen's* magnificent triple-expansion engines, by the builder, were slightly smaller than those of her consort, the *Bristol Queen*.

In this 1954 photograph she makes an elegant sight as she slows down to berth at the Barry pontoon. An affectionate and finely illustrated study of the *Cardiff Queen* by Nick James gives a full appreciation of this lovely paddler's career. Withdrawn at the end of 1966 she was scrapped in 1968 by John Cashmore Ltd., the Newport breakers, after a tragically short life by the standards of longevity set by her White Funnel predecessors.

Maritime Photo Library

The waters of the River Avon are stippled by a slight breeze and the dense foliage of Leigh Woods on the Somerset shore reach down to the water's edge. It is a truly magnificent setting for the White Funnel paddle steamer *Cardiff Queen* as she arrives at Bristol. Edwin Keen's evocative photograph was taken on 20 September 1947, towards the end of the *Cardiff Queen's* first season in the Bristol Channel.

Courtesy Bristol Museum

The White Funnel Fleet's most faithfull work-horse was the dependable *Glen Usk*. Another paddler from the Ailsa Shipbuilding Company, the *Glen Usk* (524grt) was built in 1914 and is fondly remembered as the Cardiff to Weston ferry steamer in her latter years. She was laid up in 1957 and again in 1961 and 1962, finally going the following year to Irish breakers at Passage West.

In happier post-war days – the date of this picture is 15 May 1948 – this is a reminder of the large crowds which still patronised the paddlers on a good day. By contemporary standards the *Glen Usk* did not look as elderly as the *Ravenswood* but her heavily pronounced counter stern and straight stem gave her a matronly air nonetheless!

Courtesy Bristol Museum

An engine-room shot gives us a view of her compound machinery.

Courtesy Bristol Museum

Capt. Albert Murphy poses for the camera and the gulls wait to be fed in this scene from White Funnel days in the '50s.

Photo: Sid Gray.

A heavy swell sweeps into Sand Bay. I'ts a grey day in the 1950s as the *Ravenswood* digs her stern into a roller backing out from Weston Pier. I just hope the winchman has finished coiling the stern-rope! One can almost feel, with the passengers, the sudden loss of speed as the paddles in the trough have little more than thin air to bite upon, the juddering sensation under the feet and the sound of clinking glasses and bottles in the bar below.

The *Ravenswood* is recorded to have made at least two cross-channel trips to France when stationed for three seasons (1923-25) on the south coast of England, in the days when she had an open foredeck. Having been built in 1891 she was some sixty years old when this photograph was taken.

The *Glen Gower* arriving at Brighton in June 1955. New aftermasts had been fitted to the older ships in the previous year and certainly the *Glen Gower's* profile was enhanced by this addition.

Courtesy Bristol Museum.

Penarth Dock provided a haven for the White Funnel Fleet during the winter after the move from Bristol. The photograph was taken on 2 May 1959. The *Cardiff Queen*, nearest the camera, is looking smartly turned out in preparation for her first sailings, on the Weston ferry, on 14 May. Neither the *Glen Gower*, on the wall astern, nor the *Bristol Queen*, alongside, would sail that season. The following year on 7 April the *Glen Gower* left Penarth under tow for Belgian breakers.

Grahame Farr collection.

A well photographed episode in the *Glen Usk's* latter career occurred one evening in the River Avon on a falling tide after the steamer's departure from Bristol. Some 600 trippers were returning on 30 August 1959, to Clevedon, Penarth and Cardiff when, instead of taking the Horseshoe Bend at the customary gallop, the *Glen Usk* slowed down and was pushed ashore by the swiftness of the ebb. Nothing would induce her to budge.

There was only one way out of the dilemma: all the passengers clambered over the gangways and up the muddy banks of the receding river on to the Avonmouth railway line and the Portway above. It was home to South Wales by train.

Though none of the passengers was any the worse for the experience the *Glen Usk* looked most forlorn in the gathering gloom, her decks completely deserted. Her position can be gauged from the Old Powder House downstream, a desolate stretch for the ship's crew to spend the night, and naturally there were fears for the way the *Glen Usk* would settle.

Courtesy Woodspring Museum.

This little huddle shows Mr. Brooks, the Mate on the left of the picture. Captain Neville Cotman and Mr. Keating, the Purser, are deep in conversation. An abandoned push-chair suggests that one young passenger got home well past bed-time. Mr. Smith-Cox arrived on the scene (from a dinner party that never was) to check the closing of the stoke-hold ports, they were usually left open on a voyage and would let the water in if the ship tilted. The fire-brigade came down the track from the Portway with buoyancy tanks to wedge under the port paddle box which would aid the *Glen Usk's* stability when the tide returned.

Courtesy Bristol Museum

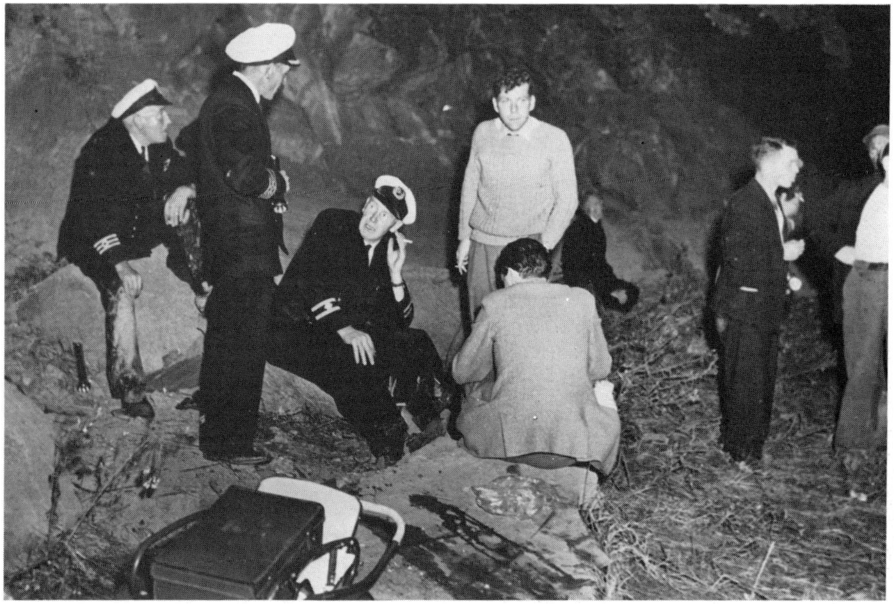

This dramatic photograph illustrates the final extent of the *Glen Usk's* list. As she settled, one of the gangways became imprisoned under the rim of the paddle box and had to be smashed with an axe.

There was nothing more to be done before the officers and crew settled down to their lonely night-long vigil. One final visitor, however, turned out to be the most welcome of them all in the person of Mr. Dawson, the resourceful Manager of the Grand Spa Hotel. Having heard of the Glen Usk's unexpected predicament and weighing up the effects of a chill, autumnal night on the health of a dispirited ship's crew, Mr. Dawson judged the situation perfectly. Immaculately attired in black coat and striped trousers, he stepped through the muddy detritus of the foreshore, with full professional aplomb, bearing his ship's provisions. In such an unlikely venue for a delivery, was a case of Navy rum ever received more unexpectedly or consumed more gratefully?

The following morning, with the assistance of two tugs, the *Glen Usk* refloated none the worse for her adventure and was back in commission after missing only two days' sailings.

Courtesy Bristol Museum.

ONE OF THE methods adopted to strengthen the commercial operation of P. & A. Campbell's ships was the innovation by Mr. Smith-Cox of regular Showboat Cruises. His position as Managing Director of the Grand Spa Hotel gave Campbells access to the fund of talented entertainers which appeared at the Spa for the cabaret at the Saturday night dinner-dances. The organiser, Reginald Williams, ensured that the artistes were contracted to appear at Hotwells, the following morning, to entertain the passengers en route to Ilfracombe aboard the Sunday Showboat Cruise.

A young Welsh singer, later to find international fame – none other than Shirley Bassey – made one of her early appearances on board a Campbells steamer. There was the Bristol recording duo Tony Fane and David Evans and later in his career I remember David Evans doing delightful monologues in his gentlemanly manner.

The greatest stir was caused by the cabaret artiste Sabrina, who was billed to appear on the Easter Sunday Showboat, taken by *Glen Usk* on that occasion. Sabrina's main attractions were amply displayed in an act which would, no doubt, fire the indignation of a modern feminist. As the dance routine reached its close-fitting climax probably not even the Captain – certainly not the passengers – were aware that the *Glen Usk* was rapidly losing steam off Foreland Point in inverse proportion to the animation on deck. It transpired that the boiler-room was vacant and that the firemen had forsaken their, admittedly, less glamorous calling to enjoy the spectacle aloft, to the extent that the *Glen Usk* eventually arrived at Ilfracombe two hours late. The lady also appeared again on the homeward trip but the punctuality, or otherwise, of the *Glen Usk's* return is not recorded.

"That fabulous Calypso man Peter Ricardo and his fascinating Dancing Girl Jobyna," aboard a White Funnel steamer.

The Showboat Cruise was fun from the very start of the day, at Hotwells, when at 8.30 the Uppington's van arrived with the amplifiers and microphones...and the piano.

Anyone who has trundled a piano on to an excursion steamer will know the hazards: at Portishead one has to take into account two flights of steps, a gravel path, a grass verge and a pair of lock gates. At Hotwells the piano was projected on to one precipitous gangway down to the pontoon and then up a narrower gangway on to the steamer – a work of art for the sweating musicians who knew that the same procedure in reverse would face them at the end of the cruise.

A tribute here to the late Tony Mockford, popular pianist with the White Funnel Players: he was never ruffled and hugely enjoyed all of the challenges which he accepted to make music on the *Bristol Queen* and the *Cardiff Queen* in those unforgettable Showboat days. His advice to the lady singers was always. "One hand for the microphone and the other for your dress!" Although in the case of the formidable Sabrina, Tony's advice was barely necessary. There was the day when the youthful Harigan and Shellard began as a duo and ended the cruise as a solo act, the other half frantically waving – to no avail – as the *Bristol Queen* drew out of Ilfracombe on her evening run back to Bristol.

They were days of tremendous fun, and the Campbells' Queens rang to the sound of laughter and applause, arriving back at Cardiff with the satisfied hum of several hundred passengers making their way up to the waiting buses at the end of Bute Street. They were magical days which many will carry in their private memories of a good day out. And at Bristol...the piano...

Why not keep the piano on board? Quite simple: the salt air played havoc with the tuning as soon as it arrived on deck and rendered it useless after a day at sea.

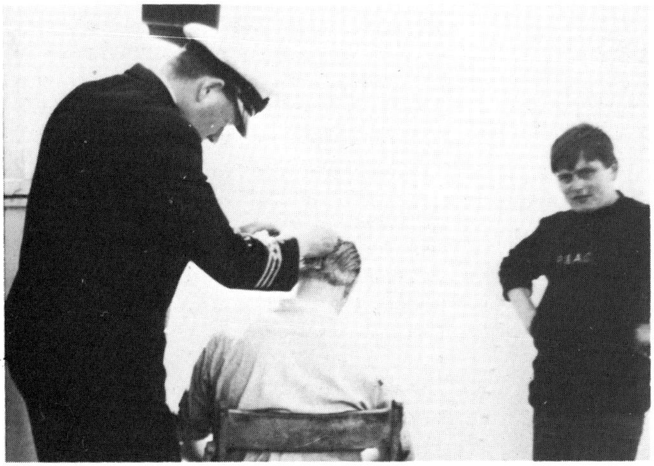

Even at the height of the season, especially when you cannot get a run ashore, certain matters have to be attended to. Identified by the colour of the funnel and the initials on the boy's jersey, a Campbell's mate gives free rein to another talent. I am reliably informed that the ship is the *Cardiff Queen*.

"The Bristol Channel is no picnic for any captain...far from it. These fickle waters require the highest possible marine experience." Thus ran an article from Campbell's in-house magazine for 1955 featuring the White Funnel captains.

From top to bottom and left to right were: Capt. Jack George (54 years experience), Master of the *Bristol Queen*; Capt. George Gunn (7 years), Master of the *Cardiff Queen* – and today still an associate of the Waverley management; Capt. A. V. Murphy (17 years), Master of the *Britannia*; Capt. J. W. Harris (6 years), Master of the *Glen Gower*; and Capt. J. A. Harris (33 years), Marine Superintendent.

P. & A. Campbell sugar bowl.

After the Second World War Campbells issued a fine series of company post cards which were sold on board the steamers. The publishers were Raphael Tuck & Sons Ltd. This classic collection, which showed the variety and strength of the post-war fleet, sold in large numbers and in recent years has become increasingly sought after by postcard collectors. Tucks also produced two minature fold-out sets, in sepia, which featured on-board views as well.

The ships featured were:

1. P. S. Bristol Queen	c. Sept. 1946	Off Battery Point, Portishead.
2. P. S. Glen Usk	c. 1956	Arriving at Weston
3. P. S. Ravenswood	c. 1948	Off Clevedon
4. P. S. Bristol Queen	c. 1948	Arriving at Weston
5. P. S. Cardiff Queen	c. 1948	Arriving at Ilfracombe
6. P. S. Glen Gower	1956	Arriving at Brighton from Bristol
7. P. S. Britannia	c. 1948	Off Brighton
8. T. S. S. Empress Queen	c. 1948	Off Brighton

Two further cards were published by Lilywhite Ltd., depicting the *Bristol Queen* arriving at Ilfracombe and the *Cardiff Queen* paddling astern from Ilfracombe Pier. Both of these attractive views were taken from the same spot and it is interesting to note the structural differences between the steamers. Curiously, this latter view of *Cardiff Queen* is the original view from the older series in which the pennants, smoke and wake were all reversed by a touch-up artist to show the paddler arriving!

A final company steamer card shows the *Cardiff Queen* off Penarth Dock in strange company. The hovercraft *SRN2 – 001* was chartered by Campbells for an experimental six-week service from Penarth to Weston, from 23 July to 30 August 1963. It was a bold innovation and deserved to be more successful than it was, however bad weather and damage at Penarth to the hovercraft's skirt meant that the experiment was not followed up locally. For the record, an inaugural run had taken special guests on 22 July. They must have been impressed by the thirteen minute flight from coast to coast. What the trippers, engulfed by clouds of sand on Weston beach, must have felt can only be contemplated.

TESTING TIMES

FROM PADDLE STEAMERS TO MOTOR VESSELS

DURING THE EARLY 1950s P. & A. Campbell still owned six paddle steamers and one turbine but, in common with all other similar operators, were finding it increasingly difficult to remain profitable. The mobility of the family motor car played havoc with other traditional forms of transport and sadly, from that moment, the days of the paddle steamers were numbered.

Although the Bristol Channel was not exempt from contemporary trends, Campbells made a crucial appointment in 1952 when Clifton Smith-Cox, a Bristol chartered accountant, joined the board. He became joint Managing Director in the following year and then sole Managing Director of P. & A. Campbell in 1955. For the next 25 years the Bristol Channel enjoyed an unbroken period of summer cruising – quite a remarkable feat in the acutely hostile financial environment and given that every other commercial operator one by one gave up the unequal struggle.

For a quarter of a century in what he himself, with some understatement, refers to as "a labour of love", Mr. Smith-Cox had difficult decisions to make: there was the removal of headquarters in 1956 from Bristol to Cardiff, the scrapping of the older paddlers as age caught up with them, the important negotiations in 1959 to place Campbells under the benign umbrella of George Nott Industries Ltd. (Townsend's Ferries) and the acquisition of motor ships which were cheaper to run.

The *Empress Queen* was withdrawn at the close of the 1951 season – the turbine in post-war days had never properly managed to fulfil the role originally mapped out for her on the cross-Channel route to France and she proved too large for coastal excursion work.

The *Ravenswood's* last season was 1954; the *Empress Queen* was sold in 1955; the *Britannia* was withdrawn at the end of 1956 and the *Glen Gower* at the end of 1957. The *Glen Usk* ran her last trips in the 1960 season. Campbells' final south coast season, in 1957, was run by the chartered motor vessel *Crested Eagle*: she ran for that year only.

It was desperately sad for Bristol Channel folk to see their old favourites disappear one by one but the future survival of the Campbell name was now dependent on scrupulous accountancy. Although Campbells' accumulated deficit was useful to George Nott Industries for taxation purposes, money still had to be found for annual overhaul and maintenance costs. The withdrawal of the older paddle steamers (all of which, except the *Glen Gower*, had also undergone the wear and tear of two wars) left the company at least a fighting chance with the two beautiful Queens.

Despite the lay-up of the *Bristol Queen* in 1959 and 1960 there followed six vintage years of the *Bristol Queen* and the *Cardiff Queen* paddling the Bristol Channel in tandem. The weather was often unspeakable but there were also some glorious spells in which a younger generation of cruising enthusiasts revelled in the lure of paddle steamer travel, loving the ships for their unique profile, their speed and that poetic thrill which only steam-driven machinery seems to inspire. Little was it realised at the time but these years graced the Bristol Channel with a spectacular final flourish of White Funnel paddle power.

The spectre of the paddle steamer as a vanishing species – not just in the Bristol Channel but throughout the United Kingdom – struck a responsive chord amongst a small band of enthusiasts and, in 1959, an Aberystwyth University Professor, Alan Robinson, founded the Paddle

Steamer Preservation Society. From very small beginnings, but fuelled by the desire to promote paddle steamer travel, active branches evolved quite naturally on the south coast of England, on the Clyde, in the Bristol Channel and in London. Their ultimate aim: to acquire, preserve and run their own paddle steamer. The outside world (even if it had known) would have looked on with wry amusement. These were early days for the society, but their time would come...

Bristol Channel enthusiasts were faintly bemused when Campbells acquired the two-funnelled motor vessel *St. Trillo* which arrived in 1963. Some were relieved that she spent much of her time cruising in North Wales waters. Of much more significance was the acquisition of the former Red Funnel ferry the *Vecta*, another motor vessel which ran in the latter part of 1965. She was renamed, more traditionally for Campbells, as the *Westward Ho* for the following and subsequent years.

The full import of *Westward Ho's* arrival became clear at the end of 1966: she was to replace the *Cardiff Queen* which was put up for sale, much to the anguish of her many admirers, and she never sailed in the Channel again.

Expectations for the 1967 season ran high and if ever there was a tide of goodwill for a ship it was for the success of the last of the White Funnel paddle steamers, the *Bristol Queen*. Sadly her fate was sealed on 26 August 1967, when her port paddle wheel was smashed, thereby hastening the inevitable, while cruising off Barry: without even the last rites of a farewell cruise, a decent valete, the *Bristol Queen's* demise was almost shockingly low-key and terribly final. Frantic efforts were made to find a role for the *Bristol Queen*, possibly as a floating restaurant in the Bristol City Docks, but they came to nothing. Harsh lessons about the financial realities of paddle steamer preservation were carefully taken to heart and stored for later application, though they were of little comfort to enthusiasts at the time: preservation just wasn't fashionable yet.

For the first time in eighty years there was no White Funnel paddler in the Bristol Channel. So deep was the sense of betrayal that some of the regular pasengers never sailed again – it was a curious attitude – but the majority quickly decided that channel trips on motor vessels were better than no trips at all.

The *St. Trillo* was retired at the end of 1969, but in the meantime Campbells had chartered the former Southampton and Isle of Wight Company's *M. V. Balmoral.* It was an inspired choice for, although the *Westward Ho* finished in 1971, the *Balmoral* carried the Campbells' flag with increasing success for the next nine seasons, on her own, in the Bristol Channel.

An indifferent summer in 1979 and the ending of the agreement with Townsend ferries that year invested the *Balmoral's* future with some uncertainty. However, for the 1980 season a new arrangement was fixed with the Landmark Trust, the Lundy Island administrators; Campbells managed the *Balmoral* for a joint enterprise known as White Funnel Steamers Ltd. It was to be Campbells' final season in the excursion trade: Lundy had designs on owning its own ship and the *Balmoral* was laid up, first in Bristol and then at Avonmouth, before being sold out of the Bristol Channel.

It was a sad ending to the story of a great company. The passing of the White Funnel Fleet evoked the deepest local regret, it was like losing an old friend. Ironically the Campbell Centenary in the Bristol Channel was the merest seven years short of its richly deserved celebration but, put in the context of the pitifully run-down state of United Kingdom shipping in general and the virtual extinction of excursion shipping in particular, for Campbells to have survived as long as they did was a magnificent feat.

Let us put the credit where it is due: had not Clifton Smith-Cox set out on his "labour of love", the last twenty-five years of the White Funnel Fleet would not have happened...and neither, quite categorically, could the extraordinary turn of events in the Bristol Channel which took place subsequently.

Campbells' first motor-vessel, the *St. Trillo*, makes an attractive sight as she plies downstream into the Horseshoe Bend in the River Avon. Speed was not one of her most noted attributes but she did have an extensive promenade deck with an ample shelter deck beneath. The *St. Trillo* (314grt) was built in 1936 on the Clyde by the Fairfield Shipbuilding & Engineering Co. The Liverpool & North Wales Steam Ship Company operated her as the *St. Silio* before renaming her in 1945. The photograph, taken on 3 April 1963, is within a month of her arrival in the Bristol Channel from Liverpool after Campbells had acquired her.

She tended to open and close the Bristol Channel season before spending the summer in her former haunts between Llandudno and the Menai Straits, a pattern which ceased with her withdrawal from service after her 1969 sailings.

Grahame Farr collection

Campbells' second motorship was the *Vecta*, built in 1938 for the Southampton – Isle of Wight service of Red Funnel Steamers. The *Vecta* made her début at Ilfracombe on 27 September 1965. The following year she adopted the White Funnel colours and became the *Westward Ho*.

With her car deck plated up and put to passenger use, the *Westward Ho* (739grt) looked smart in Campbells' livery. She was built in 1938 by John I. Thorneycroft & Co. of Southampton. There was plenty of wood in her construction (as indeed there was in the later *Balmoral* from the same company) which always enhances a ship's appearance and atmosphere.

This photograph taken on 20 June 1966, shows the motorship in Cumberland Basin, Bristol.

Her service in the Bristol Channel ended in September 1971 and she subsequently became a Manchester "floataurant". At the time of writing she is subject to conversion as an accommodation vessel at Torpoint, on the River Tamar.

Grahame Farr collection

If ever a picture tore at the very heart of shiplovers and excursionists in the Bristol Channel, it was this sight when, on 21 March 1968, the *Bristol Queen* left Cardiff under tow for the scrapyard of Van Heyghen Fréres in Belgium. With hindsight it is easy to say that it should never have been allowed to happen thus: today the outcome might have been different...perhaps. We are amazingly lucky that the motor vessel *Balmoral* and the paddle steamer Waverley have survived the British shipping depression, which slides downward with shameful regularity every year.

Out in the Channel the *Bristol Queen*, a stricken swan, glides like a ghost, past the dark lump of Penarth Head, past the distant outline of Penarth Pier – there are no excited children with seaside dreams today. Beyond, lies the faint smudge of the Somerset coast – no Minehead call, nor friendly waves from the pier at Ilfracombe as the last of the White Funnel paddle steamers passes by.

It all happened a long time ago but for those who remember the *Bristol Queen* the sound of her will not fade: the tuneful blasts from her siren coming astern out of Barry, the steamy music of her pistons in full flow, the thunder of her steel paddle-floats pounding up the River Avon on the still, evening air...the rhythms and poetry of her movement are etched on the minds of those who sailed in her.

Photo: Roy Barclay.

The *Queen of the Isles* (515grt) was surplus to the needs of the Isles of Scilly Steamship Company and Campbells chartered her for three seasons between 1968 and 1970. She had been Bristol-built by Charles Hill & Co., in 1964 – a nicely appointed ship but her lack of access forward of the bridge was a disadvantage for cruising. Nonetheless, she kept the Campbell flag flying at the North Wales piers in 1968, and on the south coast of England and in the Thames during 1969 with Calais as a planned destination. This was to be the final throw for White Funnel on the South Coast and across the English Channel; a rather sad end to a long and distinguished tradition. The *Queen of the Isles* did little work for the company in her final year.

Out on builders trials – note Hill's house flag – the *Queen of the Isles* sets off down the River Avon on 1 April 1965.

The *Balmoral*, the last regular passenger ship to fly the Campbells' house-flag, was built by Thorneycroft in 1949 as an updated version of the *Vecta (Westward Ho)*. Under White Funnel management from 1969 to 1980 she was a widely travelled ship, although what were thought, then, to be adventurous peregrinations (which included west and south coast work) proved to be a mere hors d'oeuvre to her subsequent successful round-Britain marathons!

As the memory of the great paddle steamers dimmed, the *Balmoral* began to acquire a regular clientéle – faute de mieux, perhaps, initially – but she was (and still is) wonderfully dependable with good sea-going qualities. Of her later popularity much was due to her Master, Captain Jack Wide, the last serving link with the old paddlers.

One of the special destinations was Sharpness, where the *Balmoral* called on 19 May 1979. The Gloucester branch of the World Ship Society has always given sturdy support to all Sharpness calls – a helpful example of the way in which private bodies have extended the financial viability of present-day coastal cruising.

When the Isles of Scilly Steamship Company's *Scillonian* (921grt) came up for sale in 1977 P. & A. Campbell bought her. Built in 1955 she was the third twin-screw motorship from the Thorneycroft yard to join the White Funnel Fleet. Renamed the *Devonia* she had the distinction of being the Bristol company's last acquisition – "Bristol" because, in the last years, Campbells took a lease on an office in Alliance House, Baldwin Street, from which the company operated an agency covering certain work for the parent Townsend Ferries group.

The intention was to charter out the *Devonia*, as demand required, and to use her for servicing Lundy during the winter months, a role which did not come to fruition in the latter case. She did, in fact, do oil-industry related work and some public trips in the Bristol Channel but was put up for sale with the *Balmoral* at the end of 1980.

Her subsequent career as the *Devoniun* – the nearest available change of name! – took her to Torquay, the graveyard of many larger excursion ventures. Torbay Seaways made a great success of this entry into the passenger business – to the Channel Islands – and built up such a trade that a car ferry was eventually required.

The photograph shows the *Devonia*, as yet not repainted, arriving at Ilfracombe on 29 June 1977.

Grahame Farr Collection.

A CONTINUING TRADITION

ONE SOUND WHICH the Bristol Channel never expected to hear again was the beat of paddle wheels; yet incredibly, against all odds, one ship still existed which was capable of rekindling the memories of the great steamers – the unique Clyde excursion steamer *Waverley*.

Mindful of the fate of the Campbell Queens, the Scottish members of the Paddle Steamer Preservation Society knew that when the *Waverley* was offered to them in August 1974 it was the last chance to rescue a representative paddler for posterity. By good fortune the tide was beginning to turn in favour of heritage projects and the epic story of the *Waverley's* return to active excursion work in May 1975 and her subsequent adventures, through all kinds of triumph and adversity, have captured the public imagination.

The necessity for the *Waverley* to extend her financial horizons beyond the Clyde brought her to the Bristol Channel for a long week-end during May/June 1979 and again in 1980 for a short visit . White Funnel – in what were to be the last two years of the Balmoral – generously acted as agents. The *Waverley's* presence had a profound effect on the future of local pleasure cruises, re-awakening old passions and thrilling many who had never sailed in a paddle steamer before.

Reactions were delightfully – but favourably – diverse! The older enthusiasts just shut their eyes and mentally repainted the *Waverley's* red funnels white; the younger enthusiasts revelled in the *Waverley's* new cruising waters and travelled great distances to be there; the children were captivated by the steam machinery. The less informed public persisted in the customary local fashion, to refer to her as "the Campbells" and to the population at large she was at first fairly certainly and quite erroneously recognised as the White Funnel steamer which went to Dunkirk. This deeply ingrained sense of continuity went to the very roots of the great Bristol Channel revival.

Though the White Funnel era finished at the close of the 1980 season, when the *Balmoral* was withdrawn, a group of Scottish businessmen bought the motor vessel *Shanklin* from British Rail with a view to reopening Bristol Channel services. The *Shanklin* was renamed *Prince Ivanhoe*, overhauled in Glasgow and brought down to Bristol, arriving on 1 May 1981. She was a comfortable and roomy ship, squatly good-looking and, if rather broad in the beam, well capable of adapting to the Bristol Channel.

The *Prince Ivanhoe* made a steady start but a spell of fine weather at the end of July sent passenger figures rapidly upward. Her end was as sudden as it was dramatic: on 3 August the *Prince Ivanhoe* struck an underwater obstruction off the Gower Coast on an unusually low tide and had to be beached in Port Eynon Bay. All the passengers were brought ashore. It was an unhappy end to a bold venture. The *Prince Ivanhoe's* owners, the Firth of Clyde Steam Packet Company, deserved a better reward. As if to exorcise that sad event, the *Waverley* spent an October weekend in the Bristol Channel at the close of that traumatic season.

Several very positive features became evident from this new Scottish involvement in the Bristol Channel: both ships were professionally staffed but their operators were impressed by the ancillary help which offered its services in stewarding, bar work, administration and publicity, rope-handling and sales. Volunteer assistance has always been a fundamental plank in the structure of the *Waverley's* financial success and the local branch of the Paddle

Steamer Preservation Society has rallied to the flag for her visits in every season since 1979.

With such willing back-up support and an enthusiastic Bristol Channel response to public cruises and charters, the crucial question at the headquarters of Waverley Excursions Ltd., the operating company, was: could the Bristol Channel sustain another full-time excursion ship? Was there a suitable ship anyway...?

The return to the Bristol Channel service of the motor vessel *Balmoral* in 1986 was a remarkable achievement. The Waverley organisation had spent fruitless years investigating the suitability of foreign tonnage to act as a support ship for their paddle steamer *Waverley*. Even the former General Steam Navigation Company's *Queen of the Channel* had been surveyed in Greek waters but was found to be too deep-drafted for the Bristol Channel; frequent groundings had made her hull suspect though there was a certain attraction in her cabins for overnight Highlands and Islands cruising.

All avenues seemed closed; it had become a hopeless quest. The two ships which might have provided a possible addition, the *Southsea* and the *Brading*, were resolutely unavailable. As a last resort it was agreed to look at the *Balmoral* despite reports, going back several years, that she was in a poor state.

A party of Waverley personnel discovered her at Dundee – a failed waterside restaurant and fortuitously up for sale – and her hull was surprisingly sound. One member of the inspection team was her former Bristol Channel engineer, Ken Angel, for it was rumoured that the *Balmoral's* engines were still intact: any thought of purchase hinged on this vital factor. The machinery was, indeed, still there. Would it work? At the skilled touch of practised hands the twin diesels were triumphantly coaxed into throbbing life after their long silence. The *Balmoral* story had begun a new and unexpected chapter.

It was a wonderful homecoming in Bristol. Following a lengthy and costly refit – ten times the cost of her purchase price – she left Glasgow to arrive in Bristol at 7pm on 12 April 1986. A large crowd applauded her into the lock. The *Balmoral*, "Old Faithful" of the Bristol Channel, was back, the last ship of the great White Funnel Fleet, under the Waverley flag.

Here was yet another twist to that enduring relationship between the Clyde and the Bristol Channel, with the *Balmoral* as the summer ship and the *Waverley* as a regular visitor. What would we do without the Scots!

The *Balmoral* travels extensively around Britain each season, opening up new areas for coastal cruising: she has worked in the Clyde, out of Northern Ireland and from Eire, on the south coast and up the east coast of England and Scotland but at the height of the season she is the Bristol Channel cruise ship. A motor vesel she may be but she gives pleasure to thousands in the illustrious wake of her paddle steamer forebears. The *Balmoral* truly upholds a grand tradition.

This was to be a prophetic meeting. The *Waverley* met the *Balmoral* for the first time during the afternoon of Saturday 2 June 1979, at Ilfracombe. This historic picture marks the first return to the Bristol Channel of a paddle steamer since the departure of the *Bristol Queen* in 1967. It was to be another seven years before these two ships found themselves unexpectedly under the same house-flag – the last two coastal excursion ships in the British Isles.

Photo: John Goss.

The *Prince Ivanhoe* (965grt) was built by Denny Bros. of Dumbarton, entering British Transport Commission service in 1951 and principally running between Portsmouth and Ryde. Her original name had been the *Shanklin* but she was renamed on purchase in 1980 by her new owners, the Firth of Clyde Steam Packet Company.

Her 1981 schedule took her to Steep Holm for a schools cruise on 20 May where she is seen at anchor off the island, while passengers are ferried ashore to visit the bird sanctuary. Close scrutiny reveals a ladder fixed to the front of the funnel: the Chief Engineer was in the process of adding a fine steam whistle from the old paddler *Glen Usk*. When the *Prince Ivanhoe* went aground the whistle was, sadly, taken by a souvenir hunter and every effort to trace it has failed so far.

Photo: Richard Simpson.

Like so many excursion ships before her, the motor-ship *Balmoral* starts out on her exciting new adventure – under Waverley management: Brunel's bridge has been a triumphal arch, the threshold for fresh departures, a slender frame in which to preserve the history of the Bristol Channel. The *Balmoral's* maiden trip was on 13 April 1986. In a break with Channel tradition the all-white hull and buff funnel present a most attractive sight.

Courtesy Bristol United Press.

The sense of continuity in the Bristol Channel is captured in this picture taken on the bridge of the *Waverley* during a visit to Portishead in the 1987 season. Captain David Neill, Master of the *Waverley*, has guided the fortunes of the preserved paddle steamer since the very beginning in 1975: his dedication and outstanding loyalty to this beautiful ship have earned the highest regard of all preservationists and the greatest respect of those who have sailed on her.

The group from left to right: the late Ernest Dumbleton (who sailed on the first White Funnel *Waverley*); Capt. Leo Virgo, Master of the *Cardiff Queen* in her final seasons; Capt. David Neill, Master of the *Waverley*; and Capt. George Gunn, Mate of the *Bristol Queen* and Master of the *Cardiff Queen* in the 1950s.

For Capt. Gunn the "feathered wheel" has turned full circle: his Agency covers the *Waverley* and the *Balmoral* on their Swansea and West Wales visits.

Photo: John Goss.

The *Waverley* has made it a tradition in recent years to pay an autumn week-end visit to the Bristol Channel when she returns to her Glasgow base en route from the south coast of England and the Thames.

These last calls make an end of season farewell, a re-union for steamer friends from all over Britain. It is a time for nostalgic reminiscence by the passengers while the crew, with a spring in their step, look forward to well earned leave after a strenuous summer. The date may be 13 October but there is still brilliant early evening sunshine as the *Waverley* paddles gently down the Floating Harbour for her cruise to Clevedon.

The relaxed mood of the occasion in 1989 has been superbly captured by Portishead photographer David Bunyan – the *Waverley* dressed overall and the lofty spire of St. Mary Redcliffe rising behind the Industrial Museum. This is the winter home for the *Balmoral* – seen alongside – the third member of the Paddle Steamer Preservation Society's historic preserved fleet; the paddle steamer *Kingswear Castle* is actively pleasure cruising on the River Medway.

Departure, a few moments earlier, was from the Centre, where the White Funnel steamers traditionally berthed for special occasions. Shortly she will pass Hotwells and the Mardyke where Campbell paddlers moored for the winter and Underfall Yard where they were repaired. The *Waverley* will make two short stops, like her predecessors – one at the Nova Scotia swing-bridge, another in Cumberland Basin – and latecomers will file aboard wherever the gangway goes down.

The *Waverley* passes the glass-fronted White Funnel offices in Cumberland Road, paddle steamer history enfolding her at every turn, the handsome ghosts of iron, steel and wood given shape and character – almost every one – in these very pages.

As the *Waverley* leaves the entrance locks for the muddy waters of the Avon the years fold back, over a century, to another *Waverley* from Scotland with which regular excursions in the Bristol Channel got under way. Today, the settling dusk silhouettes the trees above Rownham while, in the gloom below, the *Waverley* suddenly becomes a blaze of lights. She gathers speed in the river, the insistent, steady beat of her paddles echoing upwards into the steep sides of the Avon Gorge, past Nightingale Valley and down to Sea Mills, through the Horseshoe Bend past Chapel Pill. the Lamplighters and Pill Village, out to the open waters of the Bristol Channel.

The *Waverley*, the last of the sea-going paddle steamers, and the *Balmoral*, the last of the White Funnel Fleet, are the ultimate inheritors of a unique and continuing tradition.

Photo: David Bunyan.

ARRIVALS AND DEPARTURES

TRADITIONALLY THERE have been four different means of transferring passengers from ship to shore in the Bristol Channel: over pontoons attached to gangways; by motor-boat; on to seaside piers; and on to jetties or breakwaters which are structurally part of a harbour or dock complex. The two latter methods provide a fine distinction but one worth making.

The most important pontoons, which rose and fell with the tide, were stationed at Chepstow, Newport, Cardiff and Barry on the Welsh side and at Hotwells on the English side. The one at Chepstow fell into disrepair before the last war and all of the others succumbed to the decline in steamer services in the '50s and '60s. The neglect of these useful structures contributed towards the dwindling traffic on the Channel and helped to break established cruising patterns. As the pontoons were towed away for dismantling it was inevitable that excursion options became strictly limited: did the number of passengers justify their maintenance anyway? Such principal embarkation points were sorely missed.

Small-boat landings have always tended to be at destinations: at Lynmouth, Clovelly, Lundy and Tenby (after the pier had been dismantled). The excitement of landing by launch has always been a feature of Bristol Channel cruises, treasured by the passengers if not by the operators of the steamers. There are few regular Lundy passengers who have not enjoyed sheets of spray over the bluff bows of the motor-launch, much to the amusement of the helmsman in his oilskins and thigh-boots, to be followed by an inadvertent slip into the surf on the beach and a very soggy slog up to the village.

All this might be great fun for the passengers but for the ship's Captain with a tight schedule to keep it is a worrying business especially if the weather should turn sour during "time ashore". The pecuniary demands of local boatmen have tended to make such landings unattractive though, in fairness, the recent addition of Steep Holm landings by the Waverley and the Balmoral has been both reliable and successful.

Today, however, it is to the piers and jetties of the Channel that the steamer company must look for its safest and most predictable source of revenue. It is the hard landings and secure moorings which will ultimately assure the future of coastal excursions – not just in the Bristol Channel but all round our coasts.

The magnificent Victorian age of the pleasure pier saw to it that the Bristol Channel was was suitably provided at enough strategically placed points to help the development of steamer services. Understandably the Bristol Channel has never been able to compete with that wonderful line of piers which so consistently opened up the south coast of Britain: Bournemouth; the Isle of Wight piers at Sandown, Shanklin and Ventnor; Southsea; Brighton; Eastbourne and Hastings, an operator's delight, each with its populous summer clientéle.

Nonetheless the Channel shared locally in the development of the seaside resort with its inbuilt cruising market and there were seaside piers at Clevedon, Weston (both the Grand Pier, briefly, and Birnbeck Pier, chiefly, were used), Penarth, Minehead (up to the last war), Mumbles and Tenby (up to the last war). There were also structures at Woody Bay, which had a brief life, and at Westward Ho, which was an early example that I have yet to see a photograph of. The pier at Ilfracombe was a hybrid development, somewhere between a pleasure pier and an extension of the harbour, and is still very much in use.

Today only Penarth and Clevedon, following its amazing rejuvenation in 1989, remain open to steamer traffic.

Weston's Birnbeck Pier and Mumbles could yet, given the right initiative, be resuscitated.

Not a moment too soon, the aesthetic qualities of those wonderful monuments constructed by Victorian engineers are being rediscovered in a re-valuation, which is long overdue, but our British piers are worth preserving for many other sound reasons.

The piers of the Bristol Channel seem haunted by figures from the pages of old photograph albums, promenaders and passengers who simply tread the boards in their Sunday best or who puposefully clamber up the gangways of sleek white-funnelled steamers. Miners from the Welsh Valleys, waistcoats and watchchains, demure shop assistants in long dresses and parasols, severe nannies, chins in ruffs, school-capped youngsters...the living images of my picture postcards stride toward the pier head in a powerful and evocative sequence.

At the very beginning of our pier age the idea was that our overdressed forebears could enjoy the feeling of being at sea on a structure which didn't buck or roll, a substitute for a marine excursion which escaped the mal de mer and left Victorian dignity almost intact with nothing more than a few ruffled bonnets or, at worst, the occasional raised crinoline. If there was entertainment in the big pier pavilion and lines of animated machines to divest the owner's patrons of a few more holiday pennies, so much the better!

While the nobility took their pleasures at Bath, Harrogate or Malvern, absorbing the efficacious waters in one way or another in late Georgian times, so a vast, new city population, pent by the industrial revolution into unglamorous and often insanitary surroundings, was able to enjoy the medicinal effects of salt water and breathe the ozone at the seaside.

By the turn of the century a burgeoning network of railways, fast, efficient and relatively cheap, had placed hitherto inaccessible seaside delights within the horizons of most people. Sometimes the piers at the resorts already possessed landing stages: Birnbeck at Weston, Ilfracombe, Minehead, Mumbles, Tenby and Penarth. At Weston's Grand Pier the stage was added later. Even though steamer traffic had not originally been envisaged it soon became an extra source of revenue.

For fifty years, except during the 1914-18 war, right up to 1939, there was a golden age for the piers of Britain. There was music and entertainment, there were sideshows and arcades, all the fun of the fair and at the far end an exciting all-day sea-trip or just a "Channel dap".

In marked contrast to the piers, the small stone harbours in the Bristol Channel afford the most dramatic arrivals and departures. On the Welsh coast the White Funnel ships perched precariously at the pier-end when visiting Tenby but Porthcawl still offers an impressive berth for waiting pleasure steamers.

The delightful West Somerset harbours of Watchet and Minehead have enjoyed varying success with cruises, though the impressive revival of steam trains on the West Somerset Railway has led to an upsurge in steamer traffic. The thrill of entering the tiny harbour at Minehead underneath the tree-clad sides of North Hill is an unforgettable experience.

Channel passenger ships have also made less frequent calls at the Cornish harbours of Padstow, Newquay and St. Ives: such trips could, indeed, be placed in the marathon class.

Finally, steamers have used a variety of dockside berths, entrances, adjacent walls, locks – in fact, anywhere which sports a few well placed mooring bollards: Sharpness, Chepstow, Newport, Barry, Swansea, Pembroke and Milford Haven on the Welsh side of the Channel; Bristol, Avonmouth, Portishead, Burnham, Barnstaple and Bideford on the English coast. In recent years, as traditional moorings have become unavailable, the Waverley organisation has shown remarkable ingenuity in surveying and then opening up any safe wall with three posts and a few feet of water.

Tenby is a pleasant resort which the hyperbole of guide-books at the turn of the century elevated into the "Queen of the Welsh watering-places" and, in an early Bristol Channel Guide, "The Naples of Wales". That, at least, was the earnest view of the Town Clerk, E. Lort Stokes, Esq. who, among the other encomia of praise, pointed out that his town was "highly recommended by the Medical Profession" and, oddly, offered "Perfect Sanitation", but then, if we think about it, clean beaches have become very much a matter of the present too.

A new dimension to the town's popularity came on 9 May 1899 when the Duchess of York (later Queen Mary) opened the Victoria Pier which ran out from Castle Hill. Two and a half bowed girder spans, forerunners of the Bailey principle, rested on trestles and led to a quite substantial platform with a 300 feet landing stage. The paddle steamer *Brighton* is seen drawing away on what appears to be opening day. A rather utilitarian structure, the pier was demolished as unsafe after the Second World War. Landings thereafter were at the harbour pier or, when inconvenient tides prevailed, by small boat on to the lifeboat slipway.

Pier contractors J. & A. Mayoh completed Mumbles Pier in 1898 – it was opened on 10 May. The paddle steamer approaching is Pockett's *Brighton*, sporting a grey hull which dates the photograph before the 1905 season.

Mumbles deserves a better patronage from visiting excursion ships. Its length, 1000 feet., is quite deceptive when viewed from the shore. As at Weston's Old Pier, there is a lifeboat station attatched, but the landing stage has for the time being become unusable. In early days large numbers of passengers enjoyed a steam-train ride around the curved expanse of Swansea Bay. The railway lines are clearly visible in this photograph together with the superbly canopied entrance to the hotel on the left. The closure of the Mumbles Railway in 1960 and the decline in use of the pier for excursion work are one of the saddest aspects of this delightful Victorian village community.

There is a grand view for the bystanders at Porthcawl to greet the arrival of the *Cambria*, pre-1908.

When war broke out in 1914 the *Waverley* – here pictured by Viner, the Bath photographer – remained on the Weston to Cardiff ferry service and preserved some vestige of normality. This photograph showing the *Waverley* with round portholes (fitted by the time she was back in the Bristol Channel after being on the south coast of England in 1911) must therefore have been taken in the 1912-1916 period.

The small gaff rigged cutter between the piers is of some interest too. The large "N" on her mainsail identifies her as a Newport pilot cutter – "Cf" and "B" denoted the Cardiff and Bristol cutters. Barry was an important station for these small craft, which would set off with one or two pilots and an apprentice to meet the incoming merchant ships, often as far away as Land's End. The pilots boarded the ships bound for their home port and brought them up-channel to their journey's end. The apprentice then sailed the cutter back to base. These little craft, built for speed, were magnificent sea-boats: they had to be, to weather the storms in the Western Approaches.

The thriving coal trade brought prosperity to Penarth – as it did to all of the South Wales ports. Sadly all that remains of this fruitful era is the rather austere but imposing red-brick dock office under the shadow of Penarth Head, while the once busy Dock is now a marina.

It was, and still is, a very pleasant place to live, with substantial homes of grey stone quite reminiscent of their counterparts in Weston across the Channel. Not surprisingly Penarth became a seaside lung for Cardiff and blossomed into a resort in its own right. Every resort should have a pier if possible and in 1894 H. F. Edwards, the engineer, set to work with the contractors, James and Arthur Mayoh, to complete the structure with landing stage in 1895. Interestingly the Mayoh brothers were concurrently engaged in the building of the famous Brighton Palace Pier and Morecambe West End, while Mumbles was next on their list.

The original pier buildings at Penarth are nicely proportioned but fairly functional. The wrought iron-work of the railings, however, is a different matter: the dignified tracery of the fanlight panels is a sheer delight and still in excellent condition today, a credit to good local maintenance. The paddle steamer leaving the pier with a large crowd is the *Lady Margaret*, which was only to remain in the Bristol Channel until 1903 when she went to the Furness Railway Company's Barrow – Fleetwood service.

Alan Kittridge collection

Cardiff Pierhead was the hub of Campbells' South Wales operations. The *Waverley* waits at one of the pontoons while the *Britannia* is pulling away. On the far right of the picture the *Westonia* can be seen, representing the Red Funnel competition on the Weston ferry service.

A splendid study of P.& A. Campbell's *Albion* at the Newport pontoon. The camera has been set up on the Town Bridge which still crosses the River Usk. Most traces of the pontoon piles have disappeared, the pontoon having been withdrawn in the '60s. More recent excursions by the Waverley organisation have either used the dock entrance at the mouth of the Usk, or Blaina Wharf, downstream, on the opposite bank. The photograph was taken before 1914.

Before the new pier came into use at Chepstow, a somewhat ramshackle landing stage further down the Wye allowed passengers time ashore to view the imposing castle, Brunel's tubular bridge and the attractive little town. In this picture a totally deserted *Ravenswood* awaits the return of her passengers for whom "breaks and conveyances" were advertised as being in attendance. The romantic ruins of Tintern Abbey on the bank of the "sylvan Wye", popularised in Wordsworth's poem, were an added attraction five miles upstream, reached by rail or carriage.

An important day in the history of Chepstow was 23 May 1907, when the *Cambria* opened the new landing stage. Local photographer W. A. E. Call has captured the *Cambria* with a bow line ashore and seamen ready to throw a line from the paddle box to the pier. The hint of a wake in the middle of the Wye shows that the *Cambria* has turned further down and come up stern first to the railway bridge before going ahead into what little water there is at the pier – a somewhat delicate arrival.

Between the wars Fairfields, the shipbuilders, put a ban on the paddlers using the Wye as they considered it would interfere with activity at their slipways lower down the river.

Early photographs of steamers at Sharpness are rare. The Midland Railway ran excursions from Birmingham and the Midlands to Sharpness, connecting with the steamer there, which is possibly the purpose of the *Westward Ho's* visit in this pre-First World War postcard. "Three River" trips from Bristol (to include the Rivers Avon, Severn and the mouth of the Wye) were popular and such excursions would normally have taken the steamer to off-Sharpness rather than to call.

Sharpness, at the seaward end of the seventeen and three quarters of a mile Gloucester to Sharpness ship canal, is an interesting port where the industrial archaeologist can spend an amazingly fruitful day among the remnants of old railway sidings and hoists which exported Forest of Dean coal via the now vanished Severn railway bridge. At the Gloucester end, the basins, dry-docks, river lock and warehouses are equally fascinating and the exhibits at the newly opened Waterways Museum are quite outstanding. These are all part of the wider Bristol Channel maritime heritage to which the paddle steamers also belonged.

It is timely to remind ourselves that the steamers were just part of a transport scene which incorporated electric tramways, steam trains and the humble horse and cart, when the automobile and the motor charabanc had yet to make their mark.

The first trains came from the Gloucester district in 1892. The 1896 Channel Guide remarks that the Midland Railway, which had "always been distinguished amongst railway companies for their enterprise and liberality (sic) to the general public," ran the first trip from Birmingham on 14 July 1893. From Sharpness, on what one traveller considered "the best trip out of Birmingham, being unequalled for novelty and variety," passengers were conveyed to Weston, Lynmouth and Ilfracombe. This was typical of Campbell's enterprising approach to expanding their business.

The ancient port of Gloucester at the upper end of the Sharpness Canal was the highly unusual destination of the *Balmoral* in her first season under the Waverley Steam Navigation Company's flag. She arrived on 15 April and left on 17 April 1986: unfortunately the cruise on the intervening day, when this photograph was taken, had to be cancelled because of windy conditions. The *Balmoral* is the largest passenger ship to have visited Gloucester and she made a stirring sight in the main basin, surrounded by the historic Victorian warehouses.

This very typical Bristol scene, taken before the First World War, may not at first sight appear to be of much significance: just three vessels in Cumberland basin. However, they represent three famous Bristol companies which contributed much to the life of the port. On the left is the flagship of P.& A.Campbell's fleet of paddle steamers, *P.S. Britannia*; entering the lock on the right is the *S.S. Chicago City*, of Charles Hill's Bristol City Line. The tug, which has just eased the larger freighter through the narrow swing-bridge belongs to C.J. King & Sons, local tugowners and stevedores (now merged into the Cory Towage combine).

The two white funnels at Hotwells landing stage are the *Westward Ho* and the *Lady Ismay* astern, pictured c.1912. What appeared to be an unkempt wilderness reaching to the Clifton heights was, in fact, a mass of private glades, hanging-gardens, summer houses and secret retreats belonging to the Georgian residences behind. The sheltered slopes were a gardener's dream and to-day a stretch of the gorge below the suspension bridge must lay claim to being one of England's smallest vineyards...but it needs a sharp eye to spot it.

Photo: Harvey Barton & Son, Bristol.

There are several vantage points in the River Avon for watching and photographing ships under way. The most exciting manoeuvre occurs at the Horseshoe Bend, which a White Funnel steamer is about to negotiate. Ships can comfortably pass in the Sea Mills reach, seen here astern of the paddler and until quite recently the Sea Mills signal station either shouted or telephoned shipping movements in the lower or upper reaches to vessels in transit.

At the Horseshoe Bend the river narrows appreciably and the paddler's helmsman will hug the near bank for the deeper water, increase speed to prevent the stern being pushed ashore by the strong current, put the rudder hard over to port for the almost 180 degree swing through the bend and, probably, utter a silent prayer which, if answered, will position the ship in deep water on the far side of the river at Chapel Pill. Calling for fine judgement and a steady nerve, the Horseshoe Bend is the trickiest part of the River Avon, possibly of the whole Channel.

Again, in this pre-First World War picture, there is less mud and even a noticeable shingle strip in the area of the white marker posts by which the ships time their navigation turns. By night the beacons are lit.

Commercial postcards of steamers at Portishead are surprisingly limited. Portishead had pretensions to being a seaside resort and was a quiet, little North Somerset town of some three thousand inhabitants at the turn of the century when this picture was taken. Campbells' *Ravenswood* is just leaving the pier: the disembarking passengers have three options, to return to Bristol by train from the station which is down the steps by the flagpole, to take one of the elegant carriages to what is still called "The Village" and its shops, or to take a leisurely afternoon tea at the Royal Hotel (entrance on the left) where E. Boulter, Manageress, offered " Excellent cuisine – outing parties catered for".

The *Ravenswood's* passengers would have had the unusual sight of the wooden man o' war *Formidable* anchored at the entrance to Portishead Dock.

The *Formidable* was home to over three hundred destitute Bristol youngsters; the spartan but philanthropic régime prepared most of the boys for a life at sea, with the tender, the brigantine *Polly* in the foreground, providing sail training further afield. The *Formidable* (Capt. E. Willoughby Still) served here from 1869 to 1906 by which time a magnificent, new school – the National Nautical School – was built further down the coast at Redcliffe Bay. Note, too, the lighthouse at the entrance of Avonmouth Dock.

The artist for the Illustrated London News (10 April 1869 edition) caught the festive mood of opening day of the pier at Clevedon: the flags, the procession, and the triumphal arch made it a day to remember. Over a hundred years later the essential architecture would be miraculously unchanged; it is an area of outstanding importance and has so far escaped the crude developments of intensive planners.

The triumph for Clevedon was not, perhaps, so much the original construction of its pier in 1867-68 but rather the completion of its rebuilding in 1989 following one of the most complicated battles in which decay, lack of finance, inaction and ultimately dogged persistence all played ther part.

When Clevedon Pier was officially opened on Easter Monday, 29 March 1869, the two engineers in charge of the project – John William Grover and Richard Ward – had good reason to be proud of their creation. Stretching out into the Bristol Channel were eight spans of 100 feet., supported by the slenderest of trestles, made of lengths of Barlow rail, possibly salvaged from I. K. Brunel's South Wales Railway.

The landing stage was rebuilt and re-aligned for the 1893 season, but it was in the following summer that the pier received its crowning adornments. The Japanese pagoda flanked by two matching wrought iron pavilions were magnificent – so delicately in harmony with the graceful simplicity of the rest of the structure.

Pleasure steamers made Clevedon a regular port of call and for a hundred years the pier served cruise passengers, survived all that the inhospitable elements hurled against it and catered for the changing fashions of each passing generation. Increasingly students of architecture admired its intrinsic aesthetic beauty. Both they and the local Clevedonians were aghast when the two spans nearest to the landing stage collapsed under testing on 17 October 1970.

A dedicated band of preservationists set about raising money for its restoration. Sir Arthur and Lady Elton – the Elton family of Clevedon Court had been prime movers of the pier's original construction – became involved and enlisted the crucial support of Sir John Betjeman, whose regard for Clevedon averred it to be, with Brighton's West Pier, one of the most outstanding piers in the country.

Clevedon's shattered gem survived a local government plan for demolition, was caught up in a period of rampant inflation and afflicted by times when the gallant Pier Supporters (who never gave up hope) were confronted by financial targets which constantly seemed to gallop away from them. Then, when all seemed lost, Michael Allman, a quiet Clevedonian, decided that the sea view from his house was incomplete and single-mindedly took up the task. In 1986 the pier was dismantled and taken by barge to Portishead Dock for renovation – an incredible jig-saw of metal spars and frames. The contractors, Christiani & Nielsen, and Ernest Ireland of Bath – whose conservation skills have graced many important projects, including Lundy Village – re-assembled the structure amidst increasing local and national delight.

Amidst thrilling scenes Clevedon celebrated the re-opening of its pier on Saturday 27 May 1989, when the paddle steamer *Waverley*, specially invited for the occasion, made the first cruises. Without National Heritage funding the rebuilding could not have been accomplished but it was that undying spark of local initiative in which Clevedon can feel the greatest pride.

Dressed overall the *Waverley*, the last of the coastal paddle steamers, surges away from Clevedon Pier on 27 May 27th. Re-opening day was celebrated by two public cruises and an evening Fireworks Cruise – all well supported and undertaken in a mood of understandable euphoria! The cameraman has brilliantly caught the moment of the first departure: flags caught by the wind, port paddle-box awash as the *Waverley* digs into the turn away from the pier, Capt. David Neill on the bridge at last able to enjoy the animated scene on the crowded pier as the steamer pulls away.

Courtesy Clevedon Printing Company.

The charm of Clevedon's sea-front (unlike that of neighbouring Weston where irresistible pressures have debased the architectural integrity) is that the Victorian atmosphere has remained intact. The comfortable family houses along the promenade are flanked by the altogether grander elevations of the light and darker grey stone terraces, to which Clevedon Pier makes its own essential contribution.

The castellated toll-house – also the home of the piermaster – and the Royal Pier Hotel on the right (built in 1869) make a dignified entrance for this pre-1900 study. Happily the view is little changed today: much of the wrought iron turnstile survives, with the pavilions at the pier end to be re-instated. But note, too, the superb slot-machines, two heavily ornate period pieces of infinite charm; though why a visitor should wish to respond to the invitation to "Try Your Weight" has always struck me as an occupation of dubious pleasure, certainly on holiday.

To complete the atmosphere we are promised that the band will play on the pier, according to the notice on the gate, in the evening from 7.00 to 10.00.

There are several locations at the end of Wellington Terrace which incorporate Clevedon Pier landing stage: an attractive foreground of trees complements the arrival of a busy paddle steamer – in this case the *Cambria* – either just before or just after the first war. The *Cambria* fitted, in 1921, a small main-mast behind the after saloon to take a wireless aerial, so the picture must have been before that.

The reason for the late arrival of the Grand Pier in such a distinctive and obvious setting was, no doubt, because Weston already had another pier – Birnbeck – later known as the "Old" Pier.

Out of sight of Weston's central Promenade, hidden by Anchor Head, lies the town's finest and most under-rated architectural attraction: a pier of the utmost distinction and among the earliest dozen of such structures still surviving. Fifteen sets of wrought iron columns, centrally cross-tied and gently tapered, support the wooden decking which is underpinned by a very pleasing latticed frieze. This main span of 1,100 feet stretches out to Birnbeck Island, from which a right-angled extension of 200 feet, pointing up-channel, provided berthing for the paddle steamers.

Interestingly at the time of Birnbeck's completion in 1867 work was starting on the pier just up the coast at Clevedon. The engineer in charge was Eugenius Birch, the greatest designer of Victorian piers – fourteen to his credit, including the attractive Blackpool North and the supreme Brighton West piers.

Birch's masterpiece at Weston is best viewed from the gardens behind the entrance. At half-tide the narrowing perspective is particularly pleasing but it is at low water that Birnbeck's generous proportions are revealed in all their splendour.

In one sense the pier outlived its purpose when the steamers ceased to call: were it not for the R.N.L.I., whose activities are still an important local feature, and a sympathetic recent owner, the resourceful John Critchley, the pier would have crumbled long ago. "I am afraid it never made a profit!" Mr. Critchley ruefully observed of his eighteen year tenure. Birnbeck's future is uncertain.

The original landing stage at Birnbeck Pier could not be used at low water. The pier company therefore constructed a substantial 950 feet. low-water extension which stretched giddily downstream towards Steep Holm. It was first used in 1909 and was almost certainly the Old Pier's answer to the extension at the Grand Pier, round the corner – the competition being intense enough to produce a riposte. It was very important that the Old Pier's patrons should remain on the Island to be whirled in the chairo-planes, ducked in the water-chute and momentarily eviscerated on the switch-back than to be lured to the new, young upstart down the Promenade.

Curiously few pictures exist of this incredible structure; a number of postcards show it vaguely disappearing incompletely off the edge but only one – an aerial shot – shows its full, slender length and its enormous landing platform. In its entirety it owed much to the appearance of Clevedon, just up the coast, though the slender legs were heavily cross- strengthened, but the landing stage is pure Clevedon and mightily re-inforced.

When one looks at the paddle steamer alongside some idea can be gained of the awesome prospect facing, in this case, the Master of the *Glen Avon* in 1919 or 1920, as he nosed his way in. The vicious rocks just beneath the *Glen Avon's* bows are frightening enough. On a flood tide with five knots behind him, the Captain has no margin for error since he has but one line of approach, and if there is a following wind of any consequence his problems are compounded.

Quite undestandably Mr. Owen comments that Campbells' skippers were never happy with the arrangement, and who can blame them...The extension was dismantled in 1922. It is fascinating to observe the lengths to which rivals would go to consolidate their position in the Channel: it wasn't just a matter of having the finest ships, or the quality of the entertainments at the other end that counted but "Who can get the passengers at the times when the opposition can't ?"

This supremely evocative picture from the July section of a 1937 calendar is a summation of all that Weston must have been in those pre-war days. Visitors stroll or admire the view along the lower walk of Prince Consort Gardens, banked with hardy, salt-defying shrubs: hebe, escalonia and tamarisk. Groups of passengers, just disembarked from the steamer, make their way up Birnbeck Pier to the waiting "toast-rack" trams, of which there were six for summer running along the Front to the Sanatorium, their route since 1902.

Poignantly, even by the date on the calendar the trams had gone for good – the last runs were on the 17 April 1937 – the final, stirring clank and trundle of a system that had lost out to the motor-omnibus. The steamers no longer call and with their passing has gone, perhaps for ever, the excited cries of young, Welsh voices, the tinny rattle of buckets and spades and the satisfied scuffle of sand-filled daps, down the pier and aboard the bustling ships. Until 1960 it was still possible to see a paddler on the Weston "ferry", to linger awhile and absorb this magic view. The echo of *Glen Usk's* whistle, the splash of ropes on a still afternoon and the swash of paddle-foam catching the sunlight are childhood sights and sounds to be remembered...a white funnel, a broad, frothing wake, a paddle steamer heading for Wales – another land, another country. It fired the imagination.

You can still wander through the green railings, into Prince Consort Gardens, past the lily pond on the left and then drink in the whole extent of the middle gulf of the Bristol Channel. Sometimes, when the visibility is clear, before rain, the land contours can be spectacularly sharp as far as Foreland Point where the north Devon coast curves back into Lynmouth Bay – a distance of thirty-five miles. Across the water the smoky commerce of Cardiff is clearly defined, Penarth Head, the square bulk of Penarth Pier, Lavernock Point, down as far as the industrial chimneys at Aberthaw. It is a breathtaking panorama which invites you to sail on the waters between.

The prevailing westerlies give little chance for fog during the cruising season but early or late, the sudden changes in temperature summon up an all-embracing shroud. In the still, clammy atmosphere the fog-horn on Flat Holm fills the Channel with its mournful bellow.

The golden expanse of Weston's sandy beach is the perfect setting for a pier. The great surprise is that its construction in 1903-04 made the Grand Pier one of the very last in the noble line of British pleasure palaces. Its central position makes it the logical focus of the sea-front, a magnetic hub of all that Weston has to offer, a base for the world-famous Trapnell donkeys, a place to meet – see you by the entrance!

Like the piers at Mumbles and Penarth, Mayoh and Haley were the contractors and their conception of a huge pavilion, turretted at the four corners, seems to fill the bay from every prospect. The Grand Pier's comparative youthfulness and the updating of the pavilion after a serious fire in January 1930 have helped to make it one of the best maintained piers in the country. The sturdy decking and the immaculate paintwork at the start of each season are a credit to the Brenner family, the pier's owners. No visitor to Weston should miss the vibrance of the atmosphere or the frisson which is given by this supremely traditional seaside institution.

Yet, eighty years ago the Grand Pier had a less publicised role. Today it boasts an imposing length of 1000 feet. but, quite soon after the initial building, by 1906 a 1,500 feet. extension was added to accommodate pleasure steamers. The shallowness of the bay allowed only limited access and there were few callers. In a rare photograph the Barry Railway Company's *Devonia* is seen alongside the landing stage between August 1907 and the close of the 1908 season. The experiment was not a success and the extension was dismantled ten years later.

There are flags flying bravely for the occasion as the *Devonia* berths at the Grand Pier in brilliant sunshine and a stiff breeze flutters the name pennant, company house flag of the Barry Railway Company and the Red Ensigns on the landing stage. Who could have foreseen that this stop – no doubt a short one because of the tide – was to be repeated so rarely. The first of four calls was made by the *Ravenswood* on 16 May 1907 to open the landing stage and the Red Funnel steamers made seven calls that year. Red Funnel steamers made a further seven calls in 1908 but Campbells made none, states Mr. Owen, after which both companies refused to call.

The Red Funnel Line showed a particular interest in the development of Burnham and its link with the railway network – as can be seen, a line came right down to the pier. Other steamers called alongside when tidal conditions allowed but on this occasion the *Gwalia* is transferring passengers by small boats.

Grahame Farr collection.

The foredeck of the *Britannia* is packed solid in this photograph taken from the end of Minehead Pier. The season is c. 1902-03 and Captain Peter Campbell is on the Bridge.

Minehead was a regular destination for the *Ravenswood* between 1946 and 1954. There were several variations on the "Upper Bristol Channel Circle" theme incorporating Newport or Bristol – Clevedon – Cardiff – Penarth – Barry – Minehead. Sometimes the Bristol – Ilfracombe steamer called to provide passengers with the option to enjoy a coach trip over Exmoor or time ashore.

Unlike Watchet, Minehead's coastal cargo traffic has entirely disappeared. Even up to the late 1950's small sailing ketches and the odd schooner – with a diesel auxiliary engine by then, could be seen discharging coal for the adjacent gas-works. The small *Emma Louise* was the last to be so employed before the gas-works closed. A superb steam crane of immense antiquity effected the discharge, perched precariously, as seen here, on the edge of the abyss.

In this much earlier photograph dating from the turn of the century the schooner at the head of the harbour and the two Severn trows provide a taste of what Minehead was like in the days of sail. The trows would probably have brought Forest of Dean coal from Sharpness or Lydney; there would have been cargoes of building bricks from Bridgwater, fertiliser from Bristol for the local farms, occasionally timber from Scandinavia or roof slates from the small harbours of North Wales. Outbound cargoes included pitprops from the local forests for the South Wales coal-mines. Steamer passengers had a wealth of shipping to watch as they cruised the Bristol Channel – unlike today, most of it British: the decline of our Merchant Navy is the major unregarded scandal of our time.

The tree-lined, spacious Avenue with its comfortable, friendly market and shops is a pleasant place to visit and the climb behind the harbour-side cottages affords a superb photographic point for recording a bird's-eye view of the steamer that has brought you. One caveat, however, learned over many years is that it always takes longer to round the harbour approach than imagined, on returning to the ship. The distance is deceptive and departure time never fails to be enlivened by some quite spirited sprinting.

Tucked away inside the harbour wall at Minehead in June 1986 the *Waverley* re-embarks her passengers under the watchful eyes, on the bridge, of Bosu'n Roddy McIsaac, Channel agent Commander Tom Foden and Captain David Neill, while Purser Douglas Strachan checks the passengers at the gangway.

Watchet, the pretty little West Somerset harbour, ceased to be a port of call for the paddlers in 1929, which roughly helps to date this picture. The *Ravenswood* is the steamer arriving, little gusts of steam at the port bow exhausting from the capstan, ready for the rope to go ashore.

It was to be 1964 before Campbells called there again, but in recent years the *Waverley* and the *Balmoral* have re-awakened interest with their steam "specials" in conjunction with the West Somerset Railway.

In the face of many trials, Watchet has courageously fought to maintain regular cargo services and the re-introduction of steamer calls is a welcome addition. There is much to learn ashore about the port, its shipping and the unusual West Somerset Mineral Railway.

Courtesy Keith Thomas.

The steamer has arrived at Lynmouth for the return journey and passengers wait patiently to board the rowing boats for the long haul out to the ship.

On the sponson of the *Westward Ho* members of the crew help the passengers up to the vessel while those already aboard watch the embarkation with interest from the safety of the upper deck. The year is 1908, but the operation is a timeless one.

Lynmouth, Waiting for the Steamer.

EMBARKING AT LYNMOUTH

For a hundred years visitors to Lynmouth have enjoyed the unique charm of the Lift Railway joining the waterside village to the hotel development 500ft above at Lynton. The cliff railway opened on 7 April 1890. The project was the brainchild of the eminent Victorian publisher Sir George Newnes, proprietor of the popular paper "Tit Bits".

Sir George, who also promoted the famous Lynton & Barnstaple narrow gauge railway, must have had a passion for this unusual mode of transport for he was also the guiding force and principal shareholder in the Clifton Rocks Railway Co. Ltd., linking Hotwells with the Spa at the top. Here the first charge was fired by the Mayoress of Bristol on 7 March 1891, but it was two years before the blasting was completed and the first cars were run on 11 March 1893.

The Lynmouth railway was the longer at 900 feet. though the Clifton railway (500 feet) had four tracks to the other's two. They both used a simple system of water and gravity in which the heavier car hoisted its twin on acable. Is it too much to hope that one day an adventurous entrepreneur might reopen the Clifton Rocks Railway and restore it to its original splendour?

Courtesy Mrs. C. Coombes.

The 16th century seamen's chapel of St. Nicholas, perched high on its craggy vantage, looks down at each new arrival. Pockett's paddler, the *Brighton*, has a huge crowd on board and judging by a number of raised umbrellas on this August day in 1897, it is possibly not "Glorious Devon" today.

Lantern Hill, so-called from the ancient beacon set on the chapel roof, must have been a welcoming sight to homecoming fishermen on a murky night as the North Devon mist clamped down, a beacon of re-assurance to traders making a landfall in the inky darkness. Another comfort is the lifeboat station, whose walls carry testimony to an ongoing round of quiet and selfless service punctuated by moments of exceptional heroism in the face of the elements. The enthralling story of the rescue of 23 Spanish seamen from the steamer *Monte Gurugu*, on 13 November 1949, during a violent storm, is set unobtrusively inside the left hand door of the station. How modestly the R.N.L.I. records its triumphs!

Moving outside, a coin to hand, you find the sun shining, and the blackened tips of the steamer's funnels show reassuringly over the top of the pier. There is no perceptible movement of the masts and her name pennant hands limply or gives the occasional desultory flap to reveal a random letter or two.

A rhythmic, metallic rapping echoes from the steamer's midriff; no, it is not serious...just an engineer with a heavy spanner and hammer tightening the nuts which hold the paddle floats in place. There is a pause and, soon, he will be at work on the other wheel, suspended in space and delivering up the same hollow, cavernous echoes from his watery forge.

Tucked away in its own very private domain between Lynmouth and Combe Martin was a pier which, because of its very short life – less than four years – has always intrigued Bristol Channel habitués. The owner of the Wooda Bay estate, Benjamin Green-Lake, felt that a pier would be a useful attribute for his hotel and it was duly opened on Maundy Thursday, 1897.

P. & A. Campell called, by agreement, in the first season but in 1898, as a result, no doubt, of the paucity of the revenue returns, Mr. Green-Lake paid £300 in fees. Although Campbells still had the rights the paddle steamer Brighton did in fact call three times. Wooda Bay Pier was suddenly and totally wrecked on 29 December 1900 during a violent storm.

A contemporary view taken at low water and looking up towards the Valley of Rocks shows what an inhospitable landing it must have been at the best of times. Photographs of steamers alongside are extremely rare so it was luck to find the *Cambria* on one of her visits there. The umbrellas and the hunched figures on the sodden timbers of the pier suggest that the *Cambria* arrived in a downpour. Today, if the steamer is close in, you can still see the curved wall which led down to the landing – a stone memorial to a short-lived and long-vanished chapter in North Devon's steamer history. Wooda Bay is the pier which the pier historians have never discovered.

Photo courtesy Bristol Museum

In two photographs c.1954 the *Glen Usk* arrives with a flurry of paddle wash, letting go the anchor as she goes astern and, when steadied on her swing, takes aboard the visitors from the small boats, clustered like homing cygnets about the mother swan. The coastline sweeps round towards Westward Ho and the Taw-Torridge estuary.

Landing could be hazardous even with the slightest trace of surf. The occupants of the rowing-boat brace themselves and the waiting boatmen in thigh boots anticipate the landing, while the *Devonia* lies off. In the '20s the Lundy Island mail-boat was the *Lerina*; based at Bideford she also lies off-shore with a sail set.

The ladies with the elegant hats look splendidly incongruous as they alight from the skiff which has brought them ashore at Clovelly from the paddle steamer *Barry*, c. 1910.
Alan Kittridge collection.

One of the Lundy launches arrives at the landing-beach to re-embark passengers for the *Britannia*. The launches were kept at the island during the summer and boatmen, with assistance from the crew of the visiting steamer, would help with the landings. On shore a mobile landing stage is winched into position according to the tide, while the steamer lies off, at anchor. If the wind is easterly the landing quickly becomes choppy and, soon, impossible. There is an alternative landing (very steep, too) at Jenny's Cove on the other side of the island.

The sailing-ship days often brought grief to the shores of the Bristol Channel. One of the most spectacular strandings, however, took place in calm weather in the very middle of the Channel, on Lundy Island. The 14,000 ton battleship *H.M.S. Montagu* was engaged in wireless trials in the vicinity of Lundy when at 2.00am on the morning of 31 May 1906, the huge vessel ran ashore on Shutter Rock in a thick fog. Despite strenuous salvage efforts the *Montagu* remained firmly aground.

During the course of the summer it was an obvious attraction for the pleasure steamers. In this rare photograph the *Brighton* can be seen beyond the after part of the wreck, viewing the substantial remains. Vital equipment, including the big guns, had already been removed but the winter gales ensured that the *Montagu* did not escape from her rocky grave and she quickly broke up. The captain and navigating officer (who thought the battleship was ashore twenty miles away at Hartland Point!) were dismissed their ship – or what was left of it – at the subsequent court-martial.

Photo: Twiss Bros.

The scenic beauty of the River Torridge has always been an attraction on the rare occasions that steamers have visited Bideford in recent years. The *Bristol Queen* is on her way up to Bideford Bridge in 1954, with the railway sidings East-the-Water on the far bank. The hull of the paddler looks somewhat weather worn, suggesting an August arrival.

Courtesy Bristol Museum.

Regular visits were made to the North Cornwall port of Padstow by Pockett's paddle steamer *Brighton*. This shot from the turn of the century shows that it was no easy matter to edge a large ship into the close confines of the harbour at the mouth of the Camel estuary.

Day excursions are governed by time and distance; it was always a thrill when the north Cornish port of Padstow was the destination for the White Funnel steamer – and of course the *Waverley* has been there, too. The *Bristol Queen* makes a stirring sight, packed with passengers for the visit of 8 August 1965. The ropeman on the quay already has his bow-line coiled and is busy making ready for the spring-line amidships or the stern mooring to haul the heavier rope ashore. The visits to Padstow are highly prized by the cruising fraternity.

Courtesy Bristol Museum.

Campbells' visit to the Isles of Scilly was usually a three day affair, serviced from Penzance. The *Bristol Queen* is seen alongside the Isles of Scilly Steamship Co. Ltd.'s *Scillonian* on 19 May 1963. Fourteen years later the *Scillonian* would become part of the White Funnel Fleet as the *Devonia* but, alas, the *Bristol Queen* would be just a fond memory.

Courtesy Bristol Museum.

ACKNOWLEDGEMENTS

THIS BOOK COULD not have been written without reference to the combined knowledge of historians who have known and loved the Bristol Channel. I particularly mention the late Grahame Farr, and Sir Robert Wall, whose books have so enriched maritime studies of the Bristol Channel.

My greatest thanks go to Mr. George Owen of Swansea whose copious knowledge and detailed research into the Bristol Channel passenger steamers over seventy years has formed the basis of much that has formerly been published. I am grateful to him for the scrupulous care with which he has checked the text of this book. What has been dark, Mr. Owen has illuminated and, with the practised knowledge of an ex-Cunard Quartermaster and P&O Coxswain, he has also been kind to correct my nautical infelicities. He has been a tower of strength in my pleasant task.

I am particularly grateful to Mr. Paul Elkin of the Bristol City Museum for allowing me to view and reproduce pictures from the superb York and Keen collections.

Both Richard Clammer and Lionel Vaughan were kind enough to put their collections at my disposal.

The willing co-operation of many friends to talk, reminisce and lend me their valuable pictures has been a cornerstone of this book – as much a pleasure to them, I hope, as it certainly has been to me. Where possible I have accredited the pictures to their photographers or owners, though I beg forgiveness where their origin has become vague through copying over the years.

The Grahame Farr collection, generously offered on loan to the P.S.P.S. by his son, Michael Farr, has proved a rich source of photographs.

Unless otherwise credited, photographs are from my own collection.

Two dedicated marine engineers, Mr. Ian Muir (*P.S. Waverley* and *M.V. Balmoral*) and Mr. Ian McMillan (*M.V. Balmoral*), first implanted the idea of a book to celebrate the uniqueness of Bristol Channel cruising. In its execution, at all stages, I have enjoyed the steady and quiet encouragement of Alan Kittridge without whose judgement and good taste I could not have managed.

I owe a great debt to Mr. Clifton Smith-Cox, for kindly agreeing to write a Foreword. His resolute commitment to Bristol Channel cruising, which kept the continuity of a hundred years alive and in good heart, is a contribution of inestimable regional importance.

Finally, I thank many people for much kindness and I acknowledge their assistance: Dr. Donald Anderson, Ken Angel (Waverley S. N. Co.), Roy Barclay, Bristol United Press, David Bunyan, Richard Clammer, Jenny Coombes, Michael Farr, Ken Floyd, Sid Gray, Capt. George Gunn, Nick James, Pip Jones (Bristol Museum), Alec Lewis, Dr. Joe McKendrick, Capt. Stephen Michel (Waverley S. N. Co.), Anne Miller, Patrick Murrell, Martin Oatway, Lieut. Bryn Owen, R.N., Retd., (Welch Regiment Museum), Russell Plummer, Sidney Robinson, Alison Rosie (Scottish Record Office), Richard Simpson, Major R.P. Smith (Museum of the Royal Regt. of Wales), Keith Thomas, Peter Tippett (Sharpness Pilots), Philip Tolley, Lionel Vaughan, Woodspring Museum.

Research into the Bristol Channel and its passenger steamers is a continuing study. During the preparation of this book new information and photographs have regularly come to light. I am always happy to learn of further sources in my quest, which may be added to material already to hand for another volume.

FURTHER READING

More detailed information may be found in the following books which are relevant to the Bristol Channel:

STEAMERS

Shanklin: Ill-Fated Prince, Alan Brown.
Paddle Steamers, Richard Clammer.
West Coast Steamers, Duckworth and Langmuir.
West Country Passenger Steamers, Grahame Farr.
Balmoral Booklet, Jon Holyoak.
The Campbells of Kilmun, Iain Hope.
Cardiff Queen, Nick James.
Waverley Booklet, Fraser G. McHaffie.
The Ships That Saved an Army, Russell Plummer.
South Coast Pleasure Steamers, E.C.B. Thornton.
Bristol Channel Passenger Steamers, Robert Wall.
Paddle Wheels (P.S.P.S. Journal).
Ship Ahoy (S. Wales Branch of World Ship Society former bulletin)

PIERS

Seaside Piers, Simon H. Adamson.
Pavilions on the Sea, Cyril Bainbridge.
Clevedon Pier, Keith Mallory.

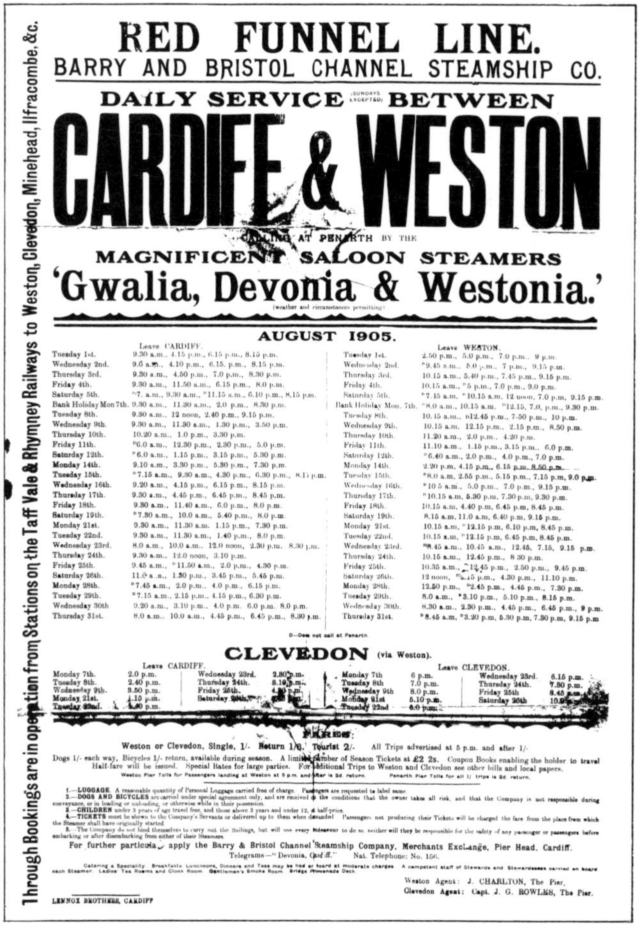

Through Bookings are in operation from Stations on the Taff Vale & Rhymney Railways to Weston, Clevedon, Minehead, Ilfracombe, &c.

RED FUNNEL LINE.

BARRY AND BRISTOL CHANNEL STEAMSHIP CO.

DAILY SERVICE (SUNDAYS EXCEPTED) BETWEEN

CARDIFF & WESTON

CALLING AT PENARTH BY THE

MAGNIFICENT SALOON STEAMERS

'Gwalia, Devonia & Westonia.'

(weather and circumstances permitting)

AUGUST 1905.

Leave CARDIFF.		Leave WESTON.	
Tuesday 1st.	9.30 a.m., 4.15 p.m., 6.15 p.m., 8.15 p.m.	Tuesday 1st.	2.50 p.m., 5.0 p.m., 7.0 p.m., 9 p.m.
Wednesday 2nd.	9.0 a.m., 4.10 p.m., 6.15. p.m., 8.15 p.m.	Wednesday 2nd.	D9.45 a.m., 5.0 p.m., 7 p.m., 9.15 p.m.
Thursday 3rd.	9.30 a.m., 4.50 p.m., 7.0 p.m., 8.30 p.m.	Thursday 3rd.	10.15 a.m., 5.40 p.m., 7.45 p.m., 9.15 p.m.
Friday 4th.	9.30 a.m., 11.50 a.m., 6.15 p.m., 8.0 p.m.	Friday 4th.	10.15 a.m., D5 p.m., 7.0 p.m., 9.0 p.m.
Saturday 5th.	D7. a.m., 9.30 a.m., D11.15 a.m., 6.10 p.m., 8.15 p.m.	Saturday 5th.	D7.15 a.m. D10.15 a.m. 12 noon, 7.0 p.m, 9.15 p.m.
Bank Holiday Mon 7th.	9.30 a.m., 11.30 a.m., 2.0 p.m., 8.30 p.m.	Bank Holiday Mon. 7th.	D8.0 a.m., 10.15 a.m. D12.15, 7.0, p.m., 9.30 p.m.
Tuesday 8th.	9.30 a.m., 12 noon, 2.40 p.m., 9.15 p.m.	Tuesday 8th.	10.15 a.m., D12.45 p.m., 7.50 p.m., 10 p.m.
Wednesday 9th.	9.30 a.m., 11.30 a.m., 1.30 p.m., 3.50 p.m.	Wednesday 9th.	10.15 a.m. 12.15 p.m., 2.15 p.m., 8.50 p.m.
Thursday 10th.	10.20 a.m., 1.0 p.m., 3.30 p.m.	Thursday 10th.	11.20 a.m., 2.0 p.m., 4.20 p.m.
Friday 11th.	D6.0 a.m., 12.30 p.m., 2.30 p.m., 5.0 p.m.	Friday 11th.	11.10 a.m., 1.15 p.m., 3.15 p.m., 6.0 p.m.
Saturday 12th.	D6.0 a.m., 1.15 p.m., 3.15 p.m., 5.30 p.m.	Saturday 12th.	D6.40 a.m., 2.0 p.m., 4.0 p.m., 7.0 p.m.
Monday 14th.	9.10 a.m., 3.30 p.m., 5.30 p.m., 7.30 p.m.	Monday 14th.	2.20 p.m, 4.15 p.m., 6.15 p.m. 8.50 p.m.
Tuesday 15th.	D7.15 a.m., 9.30 a.m., 4.30 p.m., 6.30 p.m., 8.15 p.m.	Tuesday 15th.	D8.0 a.m, 2.55 p.m., 5.15 p.m., 7.15 p.m, 9.0 p.m.
Wednesday 16th.	9.20 a.m., 4.15 p.m., 6.15 p.m., 8.15 p.m.	Wednesday 16th.	D10 5 a.m., 5.0 p.m., 7.0 p.m., 9.15 p.m.
Thursday 17th.	9.30 a.m., 4.45 p.m., 6.45 p.m., 8.45 p.m.	Thursday 17th.	D10.15 a.m, 5.30 p.m, 7.30 p.m, 9.30 p.m.
Friday 18th.	9.30 a.m., 11.40 a.m., 6.0 p.m., 8.0 p.m.	Friday 18th.	10.15 a.m, 4.40 p.m, 6.45 p.m, 8.45 p.m.
Saturday 19th.	D7.30 a.m., 10.0 a.m., 5.40 p.m., 8.0 p.m.	Saturday 19th.	8.15 a.m, 11.0 a.m, 6.40 p.m, 9.15 p.m.
Monday 21st.	9.30 a.m., 11.30 a.m. 1.15 p.m., 7.30 p.m.	Monday 21st.	10.15 a.m, D12.15 p.m, 6.10 p.m, 8.45 p.m.
Tuesday 22nd.	9.30 a.m., 11.30 a.m., 1.40 p.m., 8.0 p.m.	Tuesday 22nd.	10.15 a.m, D12.15 p.m, 6.45 p.m, 8.45 p.m.
Wednesday 23rd.	8.0 a.m., 10.0 a.m., 12.0 noon, 2.30 p.m. 8.30 p.m.	Wednesday 23rd.	D8.45 a.m., 10.45 a.m., 12.45, 7.15, 9.15 p.m.
Thursday 24th.	9.30 a.m., 12.0 noon, 3.10 p.m.	Thursday 24th.	10.15 a.m., 12.45 p.m., 8 30 p.m.
Friday 25th.	9.45 a.m., D11.50 a.m., 2.0 p.m., 4.30 p.m.	Friday 25th.	10.35 a.m., D12.45 p.m., 2.50 p.m., 9.45 p.m.
Saturday 26th.	11.0 a.m., 1.30 p.m., 3.45 p.m., 5.45 p.m.	Saturday 26th.	12 noon, D2.15 p.m., 4.30 p.m., 11.10 p.m.
Monday 28th.	D7.45 a.m., 2.0 p.m., 4.0 p.m., 6.15 p.m.	Monday 28th.	12.50 p.m., D2.45 p.m., 4.45 p.m., 7.30 p.m.
Tuesday 29th.	D7.15 a.m., 2.15 p.m., 4.15 p.m., 6.30 p.m.	Tuesday 29th.	8.0 a.m., D3.10 p.m., 5.10 p.m., 8.15 p.m.
Wednesday 30th	9.20 a.m., 3.10 p.m., 4.0 p.m. 6.0 p.m. 8.0 p.m.	Wednesday 30th.	8.30 a.m., 2.30 p.m., 4.45 p.m., 6.45 p.m., 9 p.m.
Thursday 31st.	8.0 a.m., 10.0 a.m., 4.45 p.m., 6.45 p.m., 8.30 p.m.	Thursday 31st.	D8.45 a.m, D3.20 p.m, 5.30 p.m, 7.30 p.m, 9.15 p.m

D—Does not call at Penarth.

CLEVEDON (via Weston).

Leave CARDIFF.				Leave CLEVEDON.			
Monday 7th.	2.0 p.m.	Wednesday 23rd.	2.30 p.m.	Monday 7th	6 p.m.	Wednesday 23rd.	6.15 p.m.
Tuesday 8th.	2.40 p.m.	Thursday 24th.	3.10 p.m.	Tuesday 8th	7.0 p.m.	Thursday 24th.	7.30 p.m.
Wednesday 9th.	3.50 p.m.	Friday 25th.	4.[illegible] p.m.	Wednesday 9th	8.0 p.m.	Friday 25th	8.45 p.m.
Monday 21st.	1.15 p.m.	Saturday 26th.	[illegible]	Monday 21st	5.10 p.m.	Saturday 26th	10.[illegible]
Tuesday 22nd.	[illegible].40 p.m.			Tuesday 22nd	6.0 p.m.		

[FARES]:

Weston or Clevedon, Single, 1/- Return 1/6. Tourist 2/- All Trips advertised at 5 p.m. and after 1/-

Dogs 1/- each way, Bicycles 1/- return, available during season. A limited number of Season Tickets at £2 2s. Coupon Books enabling the holder to travel Half-fare will be issued. Special Rates for large parties. For additional Trips to Weston and Clevedon see other bills and local papers.

Weston Pier Tolls for Passengers landing at Weston at 5 p.m. and after is 2d. return. Penarth Pier Tolls for all 1/ trips is 2d. return.

1.—LUGGAGE. A reasonable quantity of Personal Luggage carried free of charge. Passengers are requested to label same.
2.—DOGS AND BICYCLES are carried under special agreement only, and are received on the conditions that the owner takes all risk, and that the Company is not responsible during conveyance, or in loading or unloading, or otherwise while in their possession.
3.—CHILDREN under 3 years of age travel free, and those above 3 years and under 12, at half-price.
4.—TICKETS must be shown to the Company's Servants or delivered up to them when demanded. Passengers not producing their Tickets will be charged the fare from the place from which the Steamer shall have originally started.
5.—The Company do not bind themselves to carry out the Sailings, but will use every endeavour to do so, neither will they be responsible for the safety of any passenger or passengers before embarking or after disembarking from either of their Steamers.

For further particulars apply the Barry & Bristol Channel Steamship Company, Merchants Exchange, Pier Head, Cardiff.
Telegrams—"Devonia, Cardiff." Nat. Telephone: No. 156.

Catering a Speciality. Breakfasts, Luncheons, Dinners and Teas may be had on board at moderate charges. A competent staff of Stewards and Stewardesses carried on board each Steamer. Ladies' Tea Rooms and Cloak Room. Gentleman's Smoke Room. Bridge Promenade Deck.

Weston Agent: J. CHARLTON, The Pier.
Clevedon Agent: Capt. J. G. ROWLES, The Pier.

LENNOX BROTHERS, CARDIFF

MARITIME HERITAGE
PLACES TO VISIT

Woodspring Museum, Burlington Street, Weston. Superbly restored half-models, mirror-backed, of *Waverley (I)* and *Brighton Queen (II)*.

Weston Heritage Centre, Wadham Street, Weston. Excellent seaside display, captures resort atmosphere.

Waterways Museum, Llanthony Road, Gloucester. A mine of information on local canal network. Good video presentations. Floating exhibits, including restored steam dredger.

Bristol Museum and Art Gallery, Queens Road, Bristol. Outstanding library of local shipping photographs, five volumes on pleasure steamers alone: York Collection. Models (by appointment).

Bristol Industrial Museum, Prince's Wharf. Pleasure steamer section. Summer trips on steam tug *Mayflower.*

Bristol Maritime Heritage Centre, Gas Ferry Road. Hillhouse Collection: models and shipbuilding history. Adjacent entrance to the *Great Britain*.

Newport Museum, Steamer models (in reserve collection).

Welsh Industrial and Maritime Museum, Bute Street, Cardiff. Excellent shipping history at three sites: the main museum building at the Pier Head; a newly opened gallery at 126 Bute Street; and a special pleasure steamer building (on the site of Campbells Pier Head booking office). Railway from Queen Street Station.

Swansea Maritime Museum.

Clevedon Museum, Waterloo House. Pier history.

Watchet Museum, The Harbour. Port history.

Ilfracombe Museum, Wildersmouth. Maritime and steamer interest.

North Devon Maritime Museum, Appledore. Excellent coverage of the days of sail. Ship building. Pleasure steamers at Ilfracombe.

The Blake Museum, Bridgwater. Port History. Shipping interest.

Royal Institution of South Wales, Swansea. Material on the *Velindra* and the *Brighton*.

Front cover

The *Cambria* at speed. The curve of the bow-wave and the slurry of foam in the paddle-box give an air of urgency; the singed paintwork of the funnel and the ringed billows of smoke are sensuously real – a Campbell paddle steamer in her element!

Back cover:

The Campbell trade-mark. The abiding memory of all those who went to sea for the day in the Bristol Channel, is of slender, white funnels capped by black cowls. The funnels dominated every conceivable colour of sky and background. On sunny days they filled the mind with expectation and even on gloomy ones they uplifted the soul. Their stately poise breathed power and endowed the watcher with confidence and affection. Here the *Cardiff Queen* paddles down the Avon, past the Adam and Eve in Hung Road: a mile beyond, the Channel awaits her.

Photo: Edwin Keen, Nick James collection.